2007-2008 Assessment of the Army Research Laboratory

Army Research Laboratory Technical Assessment Board

Laboratory Assessments Board

Division on Engineering and Physical Sciences

NATIONAL RESEARCH COUNCIL
OF THE NATIONAL ACADEMIES

THE NATIONAL ACADEMIES PRESS
Washington, D.C.
www.nap.edu

THE NATIONAL ACADEMIES PRESS 500 Fifth Street, N.W. Washington, DC 20001

NOTICE: The project that is the subject of this report was approved by the Governing Board of the National Research Council, whose members are drawn from the councils of the National Academy of Sciences, the National Academy of Engineering, and the Institute of Medicine. The members of the authoring board responsible for the report were chosen for their special competences and with regard for appropriate balance.

This study was supported by Contract No. DAAD17-03-C-0017 between the National Academy of Sciences and the Army Research Laboratory. Any opinions, findings, conclusions, or recommendations expressed in this publication are those of the authors and do not necessarily reflect the views of the agency that provided support for the project.

International Standard Book Number-13: 978-0-309-14390-5
International Standard Book Number-10: 0-309-14390-X

Copies of this report are available from

Laboratory Assessments Board
Division on Engineering and Physical Sciences
National Research Council
500 Fifth Street, N.W.
Washington, DC 20001

Additional copies of this report are available from the National Academies Press, 500 Fifth Street, N.W., Lockbox 285, Washington, DC 20055; (800) 624-6242 or (202) 334-3313 (in the Washington metropolitan area); Internet, http://www.nap.edu.

Printed in the United States of America

THE NATIONAL ACADEMIES

Advisers to the Nation on Science, Engineering, and Medicine

The **National Academy of Sciences** is a private, nonprofit, self-perpetuating society of distinguished scholars engaged in scientific and engineering research, dedicated to the furtherance of science and technology and to their use for the general welfare. Upon the authority of the charter granted to it by the Congress in 1863, the Academy has a mandate that requires it to advise the federal government on scientific and technical matters. Dr. Ralph J. Cicerone is president of the National Academy of Sciences.

The **National Academy of Engineering** was established in 1964, under the charter of the National Academy of Sciences, as a parallel organization of outstanding engineers. It is autonomous in its administration and in the selection of its members, sharing with the National Academy of Sciences the responsibility for advising the federal government. The National Academy of Engineering also sponsors engineering programs aimed at meeting national needs, encourages education and research, and recognizes the superior achievements of engineers. Dr. Charles M. Vest is president of the National Academy of Engineering.

The **Institute of Medicine** was established in 1970 by the National Academy of Sciences to secure the services of eminent members of appropriate professions in the examination of policy matters pertaining to the health of the public. The Institute acts under the responsibility given to the National Academy of Sciences by its congressional charter to be an adviser to the federal government and, upon its own initiative, to identify issues of medical care, research, and education. Dr. Harvey V. Fineberg is president of the Institute of Medicine.

The **National Research Council** was organized by the National Academy of Sciences in 1916 to associate the broad community of science and technology with the Academy's purposes of furthering knowledge and advising the federal government. Functioning in accordance with general policies determined by the Academy, the Council has become the principal operating agency of both the National Academy of Sciences and the National Academy of Engineering in providing services to the government, the public, and the scientific and engineering communities. The Council is administered jointly by both Academies and the Institute of Medicine. Dr. Ralph J. Cicerone and Dr. Charles M. Vest are chair and vice chair, respectively, of the National Research Council.

www.national-academies.org

ARMY RESEARCH LABORATORY TECHNICAL ASSESSMENT BOARD

ROBERT W. BRODERSEN, University of California, Berkeley, *Chair*
DONALD B. CHAFFIN, University of Michigan
PETER M. KOGGE, University of Notre Dame
KENNETH REIFSNIDER, University of South Carolina
JOHN C. SOMMERER, Johns Hopkins University Applied Physics Laboratory
DWIGHT C. STREIT, Northrop Grumman Space Technology

Staff

JAMES P. McGEE, Director
ARUL MOZHI, Senior Program Officer
LIZA HAMILTON, Administrative Coordinator

Acknowledgment of Reviewers

This report has been reviewed in draft form by individuals chosen for their diverse perspectives and technical expertise, in accordance with procedures approved by the National Research Council's Report Review Committee. The purpose of this independent review is to provide candid and critical comments that will assist the institution in making its published report as sound as possible and to ensure that the report meets institutional standards for objectivity, evidence, and responsiveness to the study charge. The review comments and draft manuscript remain confidential to protect the integrity of the deliberative process. We wish to thank the following individuals for their review of this report:

William B. Bridges, California Institute of Technology,
Michael Dunn, Ohio State University,
David Ferguson, Boeing Company (retired),
James Glimm, State University of New York,
Thom J. Hodgson, North Carolina State University,
Mary Jane Irwin, Pennsylvania State University, and
Richard Pew, BBN Technologies.

Although the reviewers listed above have provided many constructive comments and suggestions, they were not asked to endorse the conclusions or recommendations, nor did they see the final draft of the report before its release. The review of this report was overseen by Alton D. Slay, Warrenton, Virginia. Appointed by the National Research Council, he was responsible for making certain that an independent examination of this report was carried out in accordance with institutional procedures and that all review comments were carefully considered. Responsibility for the final content of this report rests entirely with the authoring board and the institution.

Contents

APPENDIXES

Executive Summary

The charge of the Army Research Laboratory Technical Assessment Board is to provide biennial assessments of the scientific and technical quality of the Army Research Laboratory (ARL). These assessments include the development of findings and recommendations related to the quality of ARL's research, development, and analysis programs. The Board is charged to review the work only of ARL's six directorates—which excludes its reviewing two key elements of the ARL organization that manage and support basic research: the Army Research Office and the Collaborative Technology Alliances.[1] The advice provided in this report focuses on technical rather than programmatic considerations.

The Board is assisted by six National Research Council (NRC) panels, each of which focuses on the portion of the ARL program conducted by one of ARL's six directorates. When requested to do so by ARL, the Board also examines work that cuts across the directorates.

The Board has been performing assessments of ARL since 1996. The current report summarizes its findings for the 2007-2008 period, during which 95 volunteer experts in fields of science and engineering participated in the following activities: visiting ARL annually, receiving formal presentations of technical work, examining facilities, engaging in technical discussions with ARL staff, and reviewing ARL technical materials.

The Board continues to be impressed by the overall quality of ARL's technical staff and their work, as well as the relevance of their work to Army needs. The Board applauds ARL for its clear, passionate concern for the end user of its technology—the soldier in the field. While two directorates (the Human Research and Engineering Directorate and the Survivability and Lethality Analysis Directorate) have large program-support missions, there is considerable customer-support work across the directorates,

[1] Collaborative Technology Alliances are government, industry, and academic research partnerships focused on Army technologies in which the expertise resident in the private sector can be leveraged to address Army challenges.

which universally demonstrate mindfulness of the importance of transitioning technology to support immediate and near-term Army needs.

ARL staff also continue to expand their involvement with the wider scientific and engineering community. This involvement includes monitoring relevant developments elsewhere, engaging in significant collaborative work (including the Collaborative Technology Alliances), and sharing work through peer reviews (although the sensitive nature of ARL work increasingly presents challenges to such sharing).

In general, ARL is working very well within an appropriate research and development (R&D) niche and has been demonstrating significant accomplishments. Examples among many include the following:

- The development of technology for electromagnetic armor, machine translation of foreign languages, electrooptic sensors, autonomous sensing, corrugated quantum-well infrared photodetectors, robotics and unmanned air and ground vehicles, high-energy batteries, microelectromechanical systems technology for microrobotics, solid-geometry modeling computer-aided design, aircraft propulsion and structures, flexible displays, and portable biotoxin analysis;
- Research in atmospheric acoustics and radio-frequency propagation in battlefield environments, surface weather and wind modeling, auditory awareness and speech communication in battlefield environments, neuroergonomics, network science, and active stall control and active wake modeling for rotorcraft;
- The development and application of sophisticated models of soldier performance and of software to support the assessment of survivability and lethality of systems; and
- Studies to assess and improve the designs of helmets and body armor for soldiers.

ARL is increasingly addressing in proactive and creative ways challenges that require cross-directorate collaboration and is engaging in a variety of initiatives and collaborative alliances that enhance crosscutting research and development. The Board encourages ARL to continue to address several specific areas that require collaboration across ARL directorates. These include advanced computing, system-of-systems analysis, applications of neuroscience to the enhancement of soldier performance, information fusion, information security, ad hoc wireless networks, and system prototyping and model verification and validation.

ARL has been responding admirably to severe pressures to transition new technologies quickly to the field and to address simultaneously the challenging requirements of the Future Combat Systems while also maintaining its role with respect to longer-term basic research. The Board recognizes the importance of each of these types of endeavor for ARL, but it notes here the importance of basic research as a foundation for future R&D accomplishments since basic research activities may be at greater risk in the current economic environment.

ARL has been successfully addressing these significant challenges by its careful management of technical resources. Through its extensive interactions with the external academic, industrial, and government research and development communities, ARL develops opportunities to hire talented scientists, engineers, technicians, and managers. Contacts are developed through the Collaborative Technology Alliances, the Army Research Office, regular stakeholder meetings, collaborative work at the directorates, planned interaction with academic organizations, and regular recruiting activities. ARL's ability to secure needed talent would be enhanced by any administrative adjustments that improve speed and flexibility with respect to new appointments. Sufficient funding should be provided to ARL so that funding is not a constraint on managers' ability to enable the interactions of ARL staff with the scientific community through travel to professional meetings. ARL management should continue to encourage and support its staff to publish in scientific, peer-reviewed journals and proceedings.

1

Introduction

THE BIENNIAL ASSESSMENT PROCESS

This introductory chapter first describes the biennial assessment process conducted by the National Research Council's (NRC's) Army Research Laboratory Technical Assessment Board (ARLTAB). It then identifies important research areas that involve crosscutting collaboration across the Army Research Laboratory (ARL) directorates and notes the linkage between the Army Research Office (ARO) and the ARL directorates.

The charge of ARLTAB is to provide biennial assessments of the scientific and technical quality of ARL. These assessments include the development of findings and recommendations related to the quality of ARL's research, development, and analysis programs. The Board is charged to review the work of ARL's six directorates but not to review two key elements of the ARL organization that manage and support basic research: the Army Research Office and the Collaborative Technology Alliances (CTAs). Although the primary role of the Board is to provide peer assessment, it may also offer advice on related matters when requested to do so by the ARL Director; such advice focuses on technical rather than programmatic considerations. The Board is assisted by six NRC panels that focus on particular portions of the ARL program. The Board's assessments are commissioned by ARL itself rather than by one of its parent organizations.

For this assessment, ARLTAB consisted of six leading scientists and engineers whose experience collectively spans the major topics within the scope of ARL. Six panels, one for each of ARL's directorates,[1] report to the Board. Each Board member sits on a panel, five of them as panel chairs. The panels

[1] The six ARL directorates are the Computational and Information Sciences Directorate (CISD), Human Research and Engineering Directorate (HRED), Sensors and Electron Devices Directorate (SEDD), Survivability and Lethality Analysis Directorate (SLAD), Vehicle Technology Directorate (VTD), and Weapons and Materials Research Directorate (WMRD). The Board does not have a panel specifically devoted to the Army Research Office, which is another unit of ARL, but all Board

range in size from 9 to 19 members, whose expertise is tailored to the technical fields covered by the directorate(s) that they review. In total, 95 experts participated, without compensation, in the process that led to this report.

The Board and panels are appointed by the National Research Council with an eye to assembling balanced slates of experts without conflicts of interest and with balanced perspectives. The 95 experts include current and former executives and research staff from industrial research and development (R&D) laboratories, leading academic researchers, and staff from Department of Energy national laboratories and federally funded R&D centers. Twenty-six of them are members of the National Academy of Engineering, 5 are members of the National Academy of Sciences, 3 are members of the Institute of Medicine, a number have been leaders in relevant professional societies, and several are past members of organizations such as the Army Science Board and the Defense Science Board. The Board and its panels are supported by NRC staff, who interact with ARL on a continuing basis to ensure that the Board and panels receive the information that they need to carry out their assessments. Board and panel members serve for finite terms, generally 4 years, staggered so that there is regular turnover and a refreshing of viewpoints.

Biographical information on the Board members appears in Appendix B, along with a list of the members of each panel.

Preparation and Organization of This Report

The current report is the fifth biennial report of the Board. Its first biennial report was issued in 2000, and annual reviews by the Board were issued in 1996, 1997, and 1998. As with the earlier reviews, this report contains the Board's judgments about the quality of ARL's work (Chapters 2 through 7 focus on the individual directorates). The rest of this chapter explains the rich set of interactions that support those judgments.

The amount of information that is funneled to the Board, including the consensus evaluations of the recognized experts who make up the Board's panels, provides a solid foundation for a thorough peer review. This review is based on a large amount of information received from ARL and on panel interactions with ARL staff. Most of the information exchange occurs during the annual meetings convened by the respective panels at the appropriate ARL sites. Both at scheduled meetings and in less formal interactions, ARL evinces a very healthy level of information exchange and acceptance of external comments. The assessment panels engaged in many constructive interactions with ARL staff during their annual site visits in 2007 and 2008. In addition, useful collegial exchanges took place between panel members and individual ARL investigators at outside meetings as ARL staff members sought additional clarification about panel comments or questions and drew on panel members' contacts and sources of information.

Each panel meeting lasted 2½ days, during which time the panel members received a combination of overview briefings by ARL management and technical briefings by ARL staff. Prior to the meetings, some panels received extensive materials for review, including selected staff publications.

The overview briefings brought the panels up to date on ARL's long-range planning. This context-building step is needed because the panels are purposely composed mostly of people who—while experts in the technical fields covered by the directorates(s) that they review—are not engaged in work focused on Army matters. Technical briefings for the panels focused on the R&D goals, strategies, methodologies, and results of selected projects at the laboratory. Briefings were targeted toward coverage of a

panels examine how well the development of ARO and ARL are coordinated. Appendix A provides information summarizing the organization and resources of ARL and its directorates.

representative sample of each directorate's work over the 2-year assessment cycle. Briefings included poster sessions that allowed direct panelist interaction with other projects that either were not covered in the briefings or had been covered in prior years.

Ample time during both overview and technical briefings was devoted to discussion, both to clarify the relevant panel's understanding and to convey the immediate observations and understandings of individual panel members to ARL's scientists and engineers. The panels also devoted sufficient time to closed-session deliberations, during which they developed consensus findings and identified important questions or gaps in panel understanding. Those questions or gaps were discussed during follow-up sessions with ARL staff so that the panel was confident of the accuracy and completeness of its assessments. Panel members continued to refine their findings, conclusions, and recommendations during written exchanges and teleconferences among themselves after the meetings.

In addition to the insights that they gained from the panel meetings, Board members received exposure to ARL and its staff at Board meetings each winter. Also, some Board members attended the annual ARL Program Formulation Workshop in 2007 and 2008; at these workshops the ARL directorates discussed their programs with the directorates' customers and stakeholders. In addition, several panel members attended the 2007 and 2008 symposia that highlighted progress among ARL's Collaborative Technology Alliances, and selected CTA projects performed by ARL researchers were presented during panel meetings.

Assessment Criteria

Within the general framework described above, the Board developed and the panels applied detailed assessment criteria organized in the following six categories (Appendix C presents the complete set of assessment criteria):

1. *Effectiveness of interaction with the scientific and technical community*—criteria in this category relate to cognizance of and contribution to the scientific and technical community whose activities are relevant to the work performed at ARL;
2. *Impact on customers*—criteria in this category relate to cognizance of and contribution in response to the needs of the Army customers who fund and benefit from ARL R&D;
3. *Formulation of projects' goals and plans*—criteria in this category relate to the extent to which projects address ARL strategic goals and are planned effectively to achieve stated objectives;
4. *R&D methodology*—criteria in this category address the appropriateness of the hypotheses that drive the research, of the tools and methods applied to the collection and analysis of data, and of the judgments about future directions of the research;
5. *Capabilities and resources*—criteria in this category relate to whether current and projected equipment, facilities, and human resources are appropriate to achieve success of the projects; and
6. *Responsiveness to the Board's recommendations*—with respect to this criterion, the Board does not consider itself to be an oversight committee. The Board has consistently found ARL to be extremely responsive to its advice, so the criterion of responsiveness encourages discussion of the variables and contextual factors that affect ARL's implementation of responses to recommendations rather than an accounting of responses to the Board's recommendations.

During the assessment, the Board considered the following questions posed by the ARL Director:

1. Is the scientific quality of the research of comparable technical quality to that executed in leading federal, university, and/or industrial laboratories both nationally and internationally?
2. Does the research program reflect a broad understanding of the underlying science and research conducted elsewhere?
3. Does the research employ the appropriate laboratory equipment and/or numerical models?
4. Are the qualifications of the research team compatible with the research challenge?
5. Are the facilities and laboratory equipment state of the art?
6. Does the research reflect an understanding of the Army's requirement for the research or the analysis?
7. Are programs crafted to employ the appropriate mix of theory, computation, and experimentation?
8. Is the work sufficiently unique and appropriate to the ARL niche?
9. Are there especially promising projects that, with application of adequate resources, could produce outstanding results that could be transitioned ultimately to the field?

Preparation of the Report

This report represents the Board's consensus findings and recommendations, developed through deliberations that included consideration of the notes prepared by the panel members summarizing their assessments. The Board's aim with this report is to provide guidance to the ARL Director that will help ARL sustain its process of continuous improvement. To that end, the Board examined its extensive and detailed notes from the many Board, panel, and individual interactions with ARL over the 2007-2008 period. From those notes it distilled a shorter list of the main trends, opportunities, and challenges that merit attention at the level of the ARL Director. The Board used that list as the basis for this report. Specific ARL projects are used to illustrate these points in the following chapters when it is helpful to do so, but the Board did not aim to present the Director with a detailed account of 2 years' worth of interactions with bench scientists. The draft of this report was subsequently honed and reviewed according to NRC procedures before being released.

The approach to the assessment by the Board and its panels relied on the experience, technical knowledge, and expertise of its members, whose backgrounds were carefully matched to the technical areas within which the ARL activities are conducted. The Board and its panels reviewed selected examples of the standards and measurements activities and the technological research presented by ARL; it was not possible to review all ARL programs and projects exhaustively. The Board's goal was to identify and report salient examples of accomplishments and opportunities for further improvement with respect to the technical merit of the ARL work, its perceived relevance to ARL's definition of its mission, and apparent specific elements of the ARL resource infrastructure that is intended to support the technical work. Collectively, these highlighted examples for each ARL directorate are intended to portray an overall impression of the laboratory while preserving useful mention of suggestions specific to projects and programs that the Board considered to be of special note within the set of those examined. The Board applied a largely qualitative rather than quantitative approach to the assessment; it is possible that future assessments will be informed by further consideration of various analytical methods that can be applied. The assessment is currently scheduled to be repeated annually and reported biennially.

CROSSCUTTING ISSUES

The Board has regularly encouraged ARL to continue to support new interdisciplinary initiatives, including those that require collaboration across ARL directorates. The Board was provided a welcome opportunity by the ARL Director to examine ARL plans for three crosscutting Strategic Technology Initiatives (STIs): Advanced Computing, System of Systems Analysis, and Applications of Neuroscience to Enhancement of Soldier Performance. Panels examined additional crosscutting work that had previously been encouraged by the Board in the following areas: information fusion, information security, ad hoc wireless networks, and system prototyping and model verification and validation. Technical details of these crosscutting research areas are presented in the following chapters; a brief summary of the Board's impressions of these areas is presented here.

Advanced Computing

ARL's Strategic Technology Initiative in Advanced Computing gives clear indication that ARL views high-performance computing as a critical technology driven by requirements from a variety of applications, including armor and armaments, atmospheric modeling, aerodynamics, and computational biology, across multiple directorates. In addition, ARL's strategic plans include attention to petascale computing and to the investigation of software developments that will be needed to take advantage of potential applications of advanced computing. ARL's use of advanced computing for basic science is still evolving, with several new projects showing promise. This STI was at its inception when examined by the Board, and ARL is examining, appropriately, the following issues as plans are detailed and projects are implemented:

1. ARL applications drivers, both current and emerging;
2. Current ARL capabilities in simulation and modeling;
3. Opportunities for new algorithmic and software technologies to have an impact on ARL work;
4. Implications for high-performance computing requirements at ARL, including hardware, software stack, middleware libraries, and applications codes;
5. Movement of relevant high-performance applications to multicore embedded high-performance computers, reflecting a transition from processing in machine rooms to computing on the battlefield;
6. A method for verification and validation; and
7. Strategic planning issues, including building core competences, developing team structures, and seeking opportunities for leveraging across applications, domains, and directorates.

System of Systems Analysis

The Survivability and Lethality Analysis Directorate (SLAD) has continued to work on methodologies aimed at assessing the effectiveness of system of systems (SoS), which has been a continuing recommendation of the Board. However, SLAD's methodological development has focused increasingly on the System of Systems Survivability Simulation (S4), a fine-grained, event-driven simulation whose development is focused on human decision-making processes. Methodological development focusing on a complementary methodology, the Mission and Means Framework (MMF), has essentially stopped; MMF is an approach to decomposing missions and systems in order to analytically identify links between subsystems and mission performance. The Board continues to recommend strongly that SLAD add a third leg to its platform of SoS methodologies. This third methodology should provide enough fidelity to enable the meaningful study of scenarios in order to identify any major system-level impact of, for example, communications bandwidth; intelligence, surveillance, and reconnaissance; and precision weaponry without modeling fine-grain entities such as packet-level communications or details of terrain.

Developing this methodology in collaboration with an extramural team other than the one that has been developing the S4 tool would stimulate needed fresh perspectives in SoS analysis.

Applications of Neuroscience to Enhancement of Soldier Performance

The neuroscience group in the Human Research and Engineering Directorate (HRED) has responded well to new opportunities in this important arena. ARL is developing needed collaborations with the relevant research community by its proposed use of the Collaborative Technology Alliance mechanism, joining industry and academic groups. HRED's organization of the 1-day workshop that took place on May 8, 2008, was an excellent method of informing neuroscientists and cognitive scientists of the Army's needs and of quickly evaluating various research groups that could respond to a CTA announcement. The HRED staff indicated that the pending CTA announcement would focus on a few areas that seem most promising for basic research and that have clear applications to the Army's needs. ARL should develop advanced cognitive performance models to form the basis for hypotheses to be tested and to relate various neurological measurements to the prediction of human performance capabilities and mental workload.

Information Fusion

One of the most important new technology processes to emerge over the past few years is information fusion, or knowledge discovery, whereby disparate pieces of data are combined to yield higher-level knowledge, or information, that becomes actionable intelligence when presented in a sufficiently concise form and at the right time. Particularly in the Sensors and Electron Devices Directorate (SEDD) exciting developments in this area were demonstrated. The Board continues to encourage ARL to explore multidirectorate efforts to select some manageable set of problems—from sensing through the processing and presentation of information to the soldier—and to develop reasonably robust solutions for those problems that will help define the overall information fusion landscape and thus more general architectures. The Board continues to recommend crosscutting activities in this area, especially between SEDD and the Computational and Information Sciences Directorate (CISD), and especially with a close tie-in with the Network Science Division, that will form the veins for the tactical data driving such fusion.

Information Security

Information security remains an issue of great concern today in the wired computer network arena, both military and private, and it is of growing concern to the military as it moves to ad hoc networks formed from groups of warfighters. Therefore, the Board has encouraged ARL to develop crosscutting efforts in this area, especially in the establishment of testing facilities and organizations that help identify the specific challenges (both common and unique) faced by the Army and recognize when the best of commercially viable technologies provide some at-least-interim solutions. ARL's proposed creation of a new Network Sciences CTA and an Army-wide Information Assurance Center of Excellence—both addressed by CISD—seem to be appropriate moves to expand ARL's capabilities in the area of information security in significant ways. While mobile ad hoc networks (MANETs) are an important challenge, ARL should maintain global and long-term thinking with respect to traditional networks as well, since MANET-like systems will be increasingly integrated with traditional networks.

SLAD supports information assurance testing, determination of compliance with Army regulations and policies, and analysis and identification of critical system and network vulnerabilities that could potentially be exploited by an adversary, as well as development of mitigation strategies for all system and network vulnerabilities. Such efforts present opportunities to drive issues rather than to react when systems are presented to SLAD for testing and analysis, and the experience gained through testing and analysis of specific systems should be proactively leveraged to develop a methodology for overall network vulnerability assessment and to define specific metrics for evaluating performance in this area.

Ad Hoc Wireless Networks

Ad hoc networks are electronic networks with the following characteristics: the individual nodes attempting to communicate come in and out of contact with one another, the nodes can move dynamically (and thus affect which other nodes they may be in contact with), and they may encounter environmental constraints (e.g., power, bandwidth, real time, security) not present in traditional networks. Such networks, particularly wireless ones, are beginning to permeate many of ARL's projects, from sensor networks distributed over the battlefield, to dynamic intelligence networks aboard unmanned aerial vehicles, to intra- and intersoldier networks. Regarding the last of these, the extensive planning of studies by HRED on the effects of a variety of information networks on soldier performance is noted. Additionally, successful development of the ARL Blue Radio prototype by SEDD lays an excellent foundation for understanding at a deep level how the physics of radio transmission on the battlefield needs to interact with the flow of required information.[2] The Board has encouraged ARL to consider efforts to bring together the disparate groups engaged in these endeavors so that fertilization of approaches, code, and subsystems can engender progress across the board.

A particularly important ARL response in the area of ad hoc wireless networks has been the establishment of the Mobile Network Modeling Institute. The institute has a charter to work with external and internal organizations on end-to-end models of MANETs for tactical purposes before they are developed and to allow those models to guide both development and deployment activities. This institute is clearly appropriate, with the potential to develop large-scale networked radio codes strongly matched to emerging Army needs. However, unless this work is supported by a strong experimental component to validate and verify the models, there is a potential risk of falling short of ambitious goals.

System Prototyping and Model Verification and Validation

A continuing challenge for ARL is to ensure that appropriate verification and validation activities—validating that models developed during research programs actually reflect reality and verifying that the codes or systems that are supposedly constructed to match are in fact correct implementations of the models—are applied to projects whose results rely heavily on models. ARL should continue to explore carefully opportunities to exploit its high-performance computer and model resources for applications such as the following: hardware prototyping, predictive performance modeling of systems, and verification and validation of multiscale analysis and forecast models, for use in areas such as battlefield weather and the HRED-Improved Performance Research Integration Tool (IMPRINT) modeling of soldier performance in advanced hardware systems. Continued progress in this area should reduce significantly the costs of system hardware and software development and testing. The Board also continues to encourage ARL to consider ways of capturing the results of many of the field tests that it performs every year relative to such phenomena so that these results can be searched later for answers to questions not yet asked today.

---

[2] Blue Radio is a small, wireless network interface card that was designed by ARL as a demonstration platform for implementing sensor networks, particularly ones that will be placed randomly on the ground and thus must rely heavily on surface wave propagation rather than on free-space propagation.

LINKAGE BETWEEN ARMY RESEARCH LABORATORY AND ARMY RESEARCH OFFICE

The Board is not charged to review the work funded by the Army Research Office, which is an organizational entity within ARL. ARO is a significant basic research asset with a significant fraction of the total ARL basic research (6.1) budget. Considering the important role that basic research has had in the development of Army-relevant technologies and the similar high-payoff role that it could have in the future, the Board requested an opportunity to learn how the work portfolio of ARO is integrated into the activities normally reviewed by the Board. In response, ARL and ARO presented to each panel summaries of those 6.1 programs that ARO sponsors which are relevant to the ARL work reviewed by the given panel. The level of ARO collaboration varies across the directorates; in general, ARO demonstrated increasing attention to such collaboration, and the Board looks forward to continuing improvements in ARO's cognizance and support of the missions of the directorates.

2

Computational and Information Sciences Directorate

INTRODUCTION

The Computational and Information Sciences Directorate (CISD) was reviewed as a whole by the Panel on Digitization and Communications Science of the Army Research Laboratory Technical Assessment Board (ARLTAB) during August 21-23, 2007, and July 29-31, 2008. In addition, subgroups of the panel reviewed meteorology-related work on November 5, 2007, and work on high-performance computing (HPC) on May 30, 2008. The reviews consisted of overviews by directorate and division management, presentations on a subset of current projects, poster sessions during which project leads were available, and laboratory tours.

As of July 2008, CISD has grown to four research divisions: Advanced Computing and Computational Sciences Division (AC&CSD), Battlefield Environment Division (BED), Information Sciences Division (ISD), and Network Science Division (NSD). It also includes one infrastructure division, Information Technology, which serves all of ARL through its computing hardware, software, and staff. CISD is responsible for a continuing Collaborative Technology Alliance (CTA) on Communications and Networks and a continuing International Technology Alliance (ITA) on Network and Information Sciences. A new Advanced Decision Architectures CTA is co-managed by CISD with funding from the Human Research and Engineering Directorate (HRED).

CISD's expressed mission is to create, exploit, and harvest innovative technologies to enable knowledge superiority for the warfighter through advanced computing, network and communications sciences, information assurance techniques, and battlespace environment sensing and modeling. To carry out this mission, CISD performs research for the following purposes:

- To advance computational sciences and HPC technologies in support of Army systems;
- To perform atmospheric dynamics sensing and modeling for use in battlefield applications;

- To develop techniques for battlefield information fusion and processing, language translation, and autonomous agent control; and
- To develop self-configuring wireless network technologies that enable secure, scalable, energy-efficient, and survivable tactical networks.

Tables A.1 and A.2 in Appendix A respectively characterize the funding profile and the staffing profile for CISD.

CHANGES SINCE THE PREVIOUS REVIEW

Since the last documented review (for the 2005-2006 period),[1] several changes have affected CISD's research activities. The first of these was the major reorganization in 2008 that increased the number of divisions from three to four. Only BED remained unchanged. The new addition was the Network Science Division, formed from assets in the prior Computer and Communication Sciences Division and the High Performance Computing Division. This new division, NSD, grew out of the recognition that networking sciences at several levels had become key to many Army future needs, especially the mobile ad hoc networks (MANETs) expected to be found in profusion on future battlefields. NSD's charter emphasizes technologies that enhance tactical communications and networking capabilities both with warfighters and with sensor networks; methodologies to analyze, model, design, predict, and control the performance of such networks; and system architectures and algorithms to recognize and react to intrusion-detection events in such networks.

The former Computer and Communication Sciences Division was reformulated into the Information Sciences Division, with a charter primarily focused on fusing timely information from all relevant sources for the warfighter in real time.

In addition, the remaining assets of the former High Performance Computing Division were reformulated into the Advanced Computing and Computational Sciences Division, with a charter focused on using advanced computational sciences and high-performance computational resources. This division still manages or oversees two supercomputer facilities: the DoD Supercomputing Resource Center in Aberdeen, Maryland, and the Army High Performance Computing Research Center (AHPCRC), which has moved from Minnesota to California and is now directed by a team at Stanford University.

Along with this reorganization, the Communications and Networks CTA continues to advance survivable and secure information communication and processing over wireless mobile networks.

Also, the new International Technology Alliance on Network and Information Sciences was formed, involving participation from institutions in both the United States and Great Britain. The focus of this ITA is on managing end-to-end information flows in support of coalition decision making.

The three newly announced CTAs for fiscal year (FY) 2009 in Robotics, Cognition and Neuroergonomics, and Network Science will all directly relate to CISD activities. In line with Army Research Laboratory (ARL) initiatives, CISD has also established a new Mobile Network Modeling Institute and has continued investments through the DoD High Performance Computing Modernization program. Moreover, CISD has continued investments through the Small Business Innovation Research (SBIR) program to bring in new technologies from emerging high-technology companies.

[1] National Research Council, *2005-2006 Assessment of the Army Research Laboratory*, Washington, D.C.: The National Academies Press, 2007.

ACCOMPLISHMENTS AND ADVANCEMENTS

Since the 2005-2006 assessment, the Board has seen achievements in the Computational and Information Sciences Directorate in three areas: continuing advances that have been made in key projects begun in the past, new projects that have come to fruition in the past 2 years, and a combination of reorganizations and new initiatives that are focused on future needs. Each is discussed below.

Continuing Advances

The 2005-2006 assessment report documented several areas of research that have continued to demonstrate significant advances over the past 2 years:

- Machine translation of foreign languages;
- Atmospheric acoustics, radio-frequency (RF), and optical propagation in battlefield environments; and
- Modeling of surface-level weather, especially wind.

The machine translation work continues to demonstrate leadership in ways that directly aid the warfighter on today's battlefields and sets the stage to provide rapid help as needed when new fronts emerge. The advances reported in 2006 were centered on porting text and speech translation engines to laptops and personal digital assistants (PDAs) for field deployment, and a "best-of-breed" testing and evaluation procedure for conducting relevant bake-offs of emerging new Machine Translation (MT) programs from ARL and outside organizations (both industrial and academic).

The work in machine translation of foreign languages since then remains an archetype for goal-directed research with high value and frequent spin-offs. The big change since 2006 has been an increased emphasis not just on testing and evaluation but on actual translation, and doing so in the context of the kinds of work flows being experienced in the field. Thus, recent work has focused on handling not only voice but text. The goal for the latter is to help automate the processing of vast stacks of newly uncovered random documents not only by translating but by annotating them in ways that allow rapid key word and key phase searches after the fact. New languages beyond Arabic are receiving significant focus, from Urdu (in which there are at least limited amounts of training material) to languages based on Swahili or Hausa (in which there are often few or no training resources).

The technology being developed at ARL leverages directly the outstanding testing and evaluation procedures described in prior ARLTAB reviews. In particular, the use of hybrid machine translation schemes that apply multiple different MT programs to the same documents in both serial and parallel combinations seems to be yielding better results than are observed for any one approach in isolation. ARL's ability to realistically compare and fairly rank different MT programs is key to constructing, evaluating, and deploying such combinations.

The other laudable aspects of this machine translation work are the significant amount of collaboration involved and the strikingly effective ways in which this collaboration is used. Universities, government laboratories, and industry are all integral parts of the mix, ensuring that ARL understands the current state of the art and how to combine multiple techniques into real deliverables for emerging applications on the battlefield. Cooperative Research and Development Agreements (CRADAs) have been developed with multiple partners, and joint work has been performed with the National Institute of Standards and Technology.

Another area of clear and continuing outstanding progress in CISD is in selected areas of atmospheric acoustics and RF propagation in battlefield environments. The Board's 2005-2006 assessment report highlighted outstanding research in developing and evaluating acoustic propagation models that incorporate environmental effects, both natural and human-made, on acoustic signatures, and in developing remote sensing techniques for use in a range of environments, from open deserts, to rugged terrain, to urban environments, and in a variety of weather conditions. Urban environments have grown in importance to many current Army operations, and an understanding of how acoustics are propagated in such environments can provide significant tactical advantages.

Since 2006, ARL has continued this work, with growing emphasis not only on new sensor platforms such as very lightweight unmanned aerial vehicles (UAVs) and tethered aerostat balloons, but also on atmospheric effects such as temperature inversions and low-level wind jets and shears. Some of this work, such as on aerostat systems, had clearly identified new sources of noise that must be corrected for and had developed new near-Earth models to help in the analysis. The results deserve wider dissemination within the DoD, especially as the range of sensor platforms and the complexity of the environments continue to increase. ARL facilities and test ranges used in such efforts, especially at White Sands, New Mexico, remain unique and continue to contribute significantly to the relevance of the work.

A related area of continuing high-quality work is in atmospheric optical propagation as performed in the ARL Intelligent Optics Laboratory, now part of ISD. The previous report commented on both the quality of the laboratory facilities and the way in which such facilities were being used. This has continued over the past 2 years, with a focus on developing optical systems for high-energy laser directed-energy applications such as targeting, atmospheric imaging, and communications. The key feature of such work is in adapting, in real time, multiple beams focused on the same target over long distances in the atmosphere, in order to achieve effects that would be the equivalent of those from a single, coherent, higher-energy beam. Real-world demonstrations have been conducted both at relatively short range (a few kilometers on the ARL campus), and much longer range (between mountains in Hawaii).

Finally, work at constructing microscale wind models for complex terrains, especially where turbulence may be present, continues to exhibit very high quality, with attention paid to both computational efficiency (the desire to apply many such models to the battlefield) and verification against measurements from the White Sands, New Mexico, range.

New Advances

In addition to the continuing activities discussed above, several additional research thrusts have borne fruit over the past 2 years.

In terms of detection systems, continued research is leading to new mechanisms that can possibly identify the presence of particles, such as chemical and biological, in aerosols and the atmosphere. Both fluorescence spectra and optical scattering properties have been studied, with both experimental and modeling efforts. As with prior detector work that has led to actual deployments, the fluorescent spectral work is backed up by good experimental techniques and excellent test facilities, and the optical scattering modeling is leading to novel predictive techniques that have been well received in reputable scientific venues, both journals and conferences. Focus within the aerosol research program might beneficially be shifted from particle identification to volumetric presence, an area that seems to be of extreme interest in the case of biohazards. Applications resulting from such a switch might expand from the current identification to include concentration, harmfulness, and the development of removal scenarios. A second suggestion would be to see how these efforts dovetail into other ongoing efforts outside ARL, such as the Department of Defense's National Signatures Program.

Compact LIDAR systems (radar using light) have also been a focus of ARL activity that appears to be yielding potentially fieldable systems for providing tactical meteorological data in real time. The approach taken by the BED seems to be a good one. Partnering with industry through an SBIR grant has resulted in a 100 times reduction in weight of the LIDAR system—from 3,000 lb to 30 lb. Further work is focused on making the associated computations efficient enough for execution on embeddable computing systems. The reasons for focusing on this work are that embedded systems will soon have the needed computational power (through multicore), and ARL will be able to further reduce the size, weight, and power consumption of the LIDAR system.

Another new sensing effort that builds on existing expertise in atmospherics, optics, sensors, and computing is the development and experimental analysis of algorithms that use polarized thermal imaging to enhance targeting and tracking, that counter deceptive techniques such as camouflage and decoys, and that suppress background clutter and highlight the location of a target within a thermal image. This was an area suggested in the 2005-2006 assessment report. A focus on how to use a new lighter-weight sensor obtained through an SBIR collaboration may prove useful in developing complete packages for important next-generation applications such as the detection of improvised explosive devices (IEDs). By providing a more careful analysis of alternative technologies, such longer-range efforts are particularly valuable in supplementing other government-sponsored efforts that rely more on quickly deployable systems. Preliminary data indicate some potential for novel capabilities, but a more rigorous measurement program, coupled with enhanced algorithm development as well as an understanding of emerging and alternate sensors, will be needed to see this effort through to completion and handoff for system deployment.

An example of an effort in more basic science that seems poised to improve a range of Army systems focuses on increasing the understanding of atmospheric turbulence in wind, moisture, and temperature, especially near Earth's surface. This is a very important problem, especially as more and more Army activities occur in urban settings with complex wind paths. CISD is tackling the problem with the right balance of theory, modeling, and experimentation. The results should be applicable in the future for a variety of uses with respect to flying small UAVs, to plume dispersion from rotorcraft, to predicting the movement of chemical-biological clouds, and even to wildfires in both urban and wilderness terrains.

An example of an outstanding engineering effort is the development of the "Blue Radio," a small, wireless network interface card that is designed from the bottom up by ARL as a demonstration platform for implementing sensor networks, particularly ones that will be placed randomly on the ground and thus must rely heavily on surface wave propagation rather than on free-space propagation. This appears to be an excellent platform on which to build an experimental program because of its success in being deployed, its sophistication for addressing Army-specific issues (relative to commercial candidates), and the availability of local radio expertise. It also appears that this effort may represent one of the very few if not the lone remaining site for expertise in such radios in the Army.

Unfortunately, there seems to be neither a well-articulated path forward for exploiting the Blue Radio to the extent possible, nor a serious attempt to compare and contrast it with other sensor radio projects such as that developed at the University of California, Berkeley. A deliberate series of experiments based on this platform could be used to develop a methodology for validation and verification of the simulation, emulation, and theoretical efforts related to sensor radios. While this is only a specific example relevant to sensor networks, this methodology may very well form a strong base for analyzing other radio systems.

An important advantage of using a captive radio system such as the Blue Radio would be the opportunity to develop and demonstrate radio networks that also could host advanced applications; such applications could demonstrate the exploitation of such advanced concepts as cognition and trust and could measure the performance improvement of radio networks. This then could be used as motivation for further development of this radio system for deployment in wider systems.

The use of advanced computing for basic science is still evolving, with several new projects showing promise. A project to develop a code for designing microfluidic devices using large (1600 processor) supercomputers represented excellent science teamed with a deep understanding of how to leverage supercomputing. This project has potentially significant Army applications such as designing instruments to detect biological warfare agents. Another project to understand quantum dot formation is particularly relevant to night-vision sensors and is being pursued through an active collaboration with universities using state-of-the-art multiscale techniques.

Another project involves the calculation of binding affinities of protein-ligand complexes and focuses on dealing with ricin and other toxins, again clearly a relevant problem. The project demonstrates good algorithmic techniques. However, questions remain: Why was existing open-source software such as NAMD (NAnoscale Molecular Dynamics) not used?[2] Is there a plan for using the Army's supercomputing systems with such software? Leveraging such widely accepted codes should be used up to the point where deficiencies are present but should accelerate the pace of reaching at least first results.

Significant Reorganizations for the Future

There have been continued improvements both in the stability of the key CISD management team and in the directorate's reorganizations that represent changes to support a changing research portfolio.

Perhaps the single most important organizational accomplishment since 2006 in terms of the significant dividends that it should produce for the foreseeable future is the creation of the Network Science Division and the related Mobile Network Modeling Institute. According to ARL, this institute is an outgrowth of prior Board assessments which had indicated that a variety of issues associated with mobile networks, especially mobile ad hoc networks (called MANETs) in which network nodes come and go in time, had risen to be of crosscutting importance to ARL. The issues that the Board had discussed included network security, ad hoc wireless networks in particular, system prototyping, and model validation and verification.

The structure of the new NSD focuses on three levels of the problem: tactical network assurance, networking sciences development, and the sustaining of base network assurance. This focus represents an excellent capability, if executed well, for addressing the issues of "I can't define the problem precisely" in networks at all scales, for integrating the efforts of engineers and scientists (especially mathematicians), for starting solid validation and verification efforts for new network technologies and applications that will significantly simplify downstream system deployments resulting from divisional research, and for developing serious transition plans for such technologies. This approach should be a mechanism to allow ARL to track and evaluate in very fundamental ways new technologies that come out of both the academic and commercial worlds (such as the entry of search engine firms into the cellular telephone and cellular telephone applications arena).

The first class of problems on which NSD focuses addresses lower-level issues associated with tactical and battlefield wireless networks, from signal processing to intrusion detection, and places under one roof several of the projects that had been scattered across other ARL directorates in the past. Multiple quick-reaction laboratories have been folded into this division, which should aid in meaningful early

[2] NAMD is an open-source parallel molecular dynamics code designed for high-performance simulation of large biomolecular systems. NAMD can simulate the movement of proteins with millions of atoms, making it the world's fastest parallel molecular dynamics program.

evaluation of new networking infrastructure, and perhaps even more important, of how new applications may play on top of such networks. This capability should enhance ARL's overall laboratory facilities significantly.

The second class of problems that NSD focuses on appears to be still in its formative stages but has as its goals a much higher level understanding of networks and networking in general and the management of a variety of collaborative ventures in the networking area. The latter effort includes the existing Communications and Networks CTA and the new Network and Information Sciences ITA. The proposed creation of a new Network Sciences CTA and that of an Army-wide Information Assurance Center of Excellence seem to be appropriate moves to expand ARL's capabilities in significant ways.

The third class of problems on which NSD focuses picks up on continuing issues of security for more classical networks, as was practiced by personnel in the prior High Performance Computing Division for systems across the Army and DoD. What is especially important here is that this new organization inserts some ability to step back from the day-to-day problems of intrusion detection and think a bit more globally and longer term, at a time when more and more MANET-like systems are integrated among traditional networks.

The Mobile Network Modeling Institute works with external and internal organizations on end-to-end models of MANETs for tactical purposes before they are developed and to allow those models to guide both the development and deployment activities. These models range from environmental components (these simulate transmission characteristics in battlefield environments under a variety of weather conditions and terrains), to components relating to the signal processing needed in the network nodes for shaping transmissions and receiving them, to the software protocol layers that manage communications, to the applications layers that use the communications paths. It is expected that such models would be hosted on Army supercomputing facilities, with an intent to archive the data produced using structured format (XMDF, or eXtensible Model Data Format) for validation and later use.

This institute is clearly appropriate, with the potential to develop large-scale networked radio codes strongly matched to emerging Army needs. However, there is a potential danger of the work of the institute's being too ambitious and oversold, especially if there is not a strong experimental component to validate and verify the models. The development of high-performance computing codes by themselves is no substitute for good physics to start with, for example in ensuring adequate terrain models that can produce the accuracy and precision needed in the propagation models. Building a connection with other Army sites such as Yuma Proving Ground, Arizona; Fort Irwin, California; Fort Dix, New Jersey; and others may provide valuable sites for such experimental validation in different surface environments.

A related concern involves how some of the early projects associated with this institute, discussed in detail later, balance good academics with Army needs, particularly when optimization of some aspect of a project is part of the effort. The Blue Radio, in particular, seems to be an ideal candidate for building and then validating models of sensor networks that can be verified experimentally, but CISD did not evince any explicit recognition of such an opportunity. Also, before large-scale archiving of modeled data begins, serious thought must be given to how to correlate these data with experimental data, and perhaps how future, yet-undefined projects may want to leverage the archived data for fast development of new systems or quick response in order to determine how a deployed system might perform in a new environment.

The other major reorganization within CISD was the conversion of the former High Performance Computing Division into the new Advanced Computing and Computational Sciences Division. The revised mission of the division is to advance computational sciences and HPC technologies in support of Army systems. Most of the division, however, continues as in the past to support the computational infrastructure for ARL, particularly the DoD Major Shared Resource Center and the Army High Performance Computing Research Center—both state-of-the-art supercomputing facilities. Only part of one branch, the Computational Sciences and Engineering Branch (CSEB), performs research that falls within the scope of the Board's assessment. However, the research directions of this group have been broadened, specifically to include a focus on how to get supercomputing levels of computational power into multicore embedded systems that can be positioned closer to where they are needed. In a real sense, this reflects the view that, for the Army, the machine room of the future will be the battlefield and not just a large, air-conditioned center in the United States proper. In particular, the following moves are commendable: considering how to get petaflops in a truck (a petaflop being the level of performance for the fastest machines today); examining how to leverage the emergence of new and specialized computational engines such as the multicore microprocessor chips that are becoming ubiquitous in everything from servers to laptops and PDAs, graphics and game processor chips with extraordinary computational capability that can be used for functions other than graphics, and field-programmable arrays (semiconductor devices that can be configured after manufacturing) for specific Army applications; and beginning to examine the use of such capabilities in basic science projects to explore alternative technologies from the nanoscale or biological realms. A small effort has been started to establish an asymmetric computing center to explore the use of many of these nontraditional computing architectures for specific Army applications. This should be encouraged.

OPPORTUNITIES AND CHALLENGES

Systems Engineering

As noted in ARLTAB assessments in prior years, a significant challenge remains in ensuring that even in relatively basic research programs sufficient consideration is given to questions about how potential systems that might be developed out of such research could be deployed and used in real Army scenarios. The Board continues to suggest that a small amount of systems engineering early in many programs could help avoid paths that, even if successful, would be difficult to deploy in those systems that require real-time responses; conversely, this same effort would provide insight into alternatives that would mesh much better with practice. The same systems engineering focus would also help early on to compare research goals with expected roadmaps for established technologies and would help prepare realistic statements of the potential gains from the new technologies being researched.

One example of current research areas where such a focus might be valuable is the deployment of sensor networks, especially for chemical and biological agents where now a single detector is deployed. A corresponding determination of what is needed in terms of additional computational or network support is essential to achieving a viable detection system. One particularly strong project in this arena (on microfluidic sensor design) was being done by university scientists, and while they clearly understood the potential applications to Army missions, there was little or no thought expressed as to how, if the project is successful, transitions to deployable systems might take place. Such systems (or systems of systems) would integrate many functions, including sensing, analysis of responses, and communication of information.

The use of polarized thermal imagers is another example in which the desire exists to transition something to Army use, but to do so more quickly it would help to extrapolate potential system requirements from a suite of possible deployed configurations. Such requirements may prove invaluable in specifying appropriate sensor detection characteristics and in developing an understanding of the kinds of outputs that need to be generated by the associated processing system.

Similarly, projects that attempt to provide autonomous navigation for small robots in an urban setting where the Global Positioning System may not be available are clearly of significant value to soldiers in terms of reducing operator workload. However, simply trying to use existing computationally expensive image-recognition algorithms for building recognition may not be feasible for small mobile platforms, especially given the potential size of the required three-dimensional image database and accounting for the effects of battle damage on buildings. Some systems engineering to bound the amount of potentially available computational resources and suggest hybrid methods of navigation (using compasses, simple inertial navigation devices, and simpler image programs that construct line segments, vanishing points, and other geometric features) may in fact provide more direct and implementable solution plans. This would be true particularly if and when packs of multiple robots of this type were to be employed, and tasks such as sweep and survey in mass became important.

Some of the start-up efforts within the NSD in conjunction with the Mobile Network Modeling Institute may invoke similar concerns. An attempt to develop a science of networks at multiple levels is proceeding in a reasonable fashion mathematically, but it may be more significantly advantaged by the inclusion of a greater focus on whole systems and on overall network performance as seen by the end user in real environments—especially for those cases where the user is a soldier and the environment is a battlefield. This relates to the question of how to balance good academics with Army needs, particularly when attempts are made to optimize without a solid estimate of what the optimization will buy and a clear understanding of whether the fundamentals are well enough known at the current time to justify an optimization study. In particular, it may be beneficial to have some hardware experiments that run side by side with the theoretical work to demonstrate the applicability of the latter. One such collaboration that may pay exceptional dividends is to model sensor networks enabled by the Blue Radio project under way within CISD. Such collaboration might result in somewhat fewer narrowly focused publications, but it would produce results that would be making a more important contribution to the research in mobile ad hoc wireless networks that will be used in the future by the Army. Also, as discussed earlier, some serious system engineering thought up front as to how any archived data resulting from the institute's modeling or experimental data might be used for future problem solving, and subsequent use of that insight to help organize the archiving effort properly, might provide a very beneficial long-term resource to ARL.

Validation and Verification

Another general area addressed in prior ARLTAB reports that still remains a challenge is the testing and evaluation of experimentally driven programs and the validation and verification of models—validating that models developed during research programs actually reflect reality and verifying that codes or systems that are supposedly constructed to match are in fact correct implementations of the models. There are research areas such as machine translation in which these activities are central to the research process and others with apparently little such focus. In other research areas, particularly in projects involving complex computations for basic science, there is still a tendency to developing stand-alone codes without any clearly articulated approach to ensure that both the algorithm modeling the physics and the implementation of that algorithm are correct. A tendency to believe the machine is evident and needs to be avoided by formal verification. Further, very often these codes are in areas where the community as a whole does have standard open-source codes, such as NAMD for molecular dynamics, and those codes have already been adapted for execution on supercomputers, with which the Army is well equipped. The use of such codes outright for the computations and for the verification of new codes through careful side-by-side comparisons is warranted. Toward standardizing its practices, CISD should examine the methods by which other government and industry laboratories perform effective and efficient verification and validation.

Closely related to this issue is the challenge of obtaining comprehensive data sets to validate models completely. In many areas, such as weather or atmospherics, ARL's facilities for gathering relevant data are first rate. However, in other newer areas, such as chemical or biological agent detection, complete data sets may not be immediately available at ARL facilities, and alternative mechanisms or partnerships for gaining access to such data should be sought. Some of the newer laboratories such as the Wireless Emulation Laboratory are moving to a largely simulated environment, but without a strong plan to validate, at least periodically, such simulations against the real world, the results emerging from such facilities may not have the grounding in reality to make the results applicable to real Army problems.

Another emerging general challenge that is appearing in many programs is an increasing need to perform sophisticated analyses on experimental data. Such analyses involve both classical statistical computation and, perhaps more importantly, information extraction from large and often unstructured data sets. Data mining has emerged in the commercial world as key for applications ranging from determining personalized online purchase preferences to performing portfolio analyses. Similar techniques will become of increasing importance for CISD-relevant applications ranging from sensor network data analysis to prognostication and prediction of the dynamic health of platforms ranging from vehicles to aircraft. These data-mining techniques will also become important in looking through reams of multidimensional experimental data to develop and validate new detection algorithms and to analyze massive intelligence data sets for the detection of potential terrorist activities from non-physically-based data. The results from the models emerging from the Mobile Network Modeling Institute are examples of such data sets with potentially long-term value, but for which the schemas used for archiving and then retrieving them will make all the difference later on as to their ultimate usefulness.

Growing in-house expertise in such areas should provide ARL with opportunities for both off-line and online system implementations with extraordinary increases in autonomy and robustness. If done properly, such implementations will blend in seamlessly with more traditional numeric-oriented signal processing to produce intelligent and agile real-time control loops for a wide spectrum of future Army systems.

A closely related suite of problems of increasing importance to the Army lies in the ability to accumulate, analyze, understand, and efficiently process human and electronic intelligence about relationships between individuals and organizations in an asymmetric battlespace. This capability was referred to as " information fusion" in the 2005-2006 assessment report. CISD has recognized the importance of networks in general by the establishment of the new NSD, and it is clear that several new projects within ISD are oriented toward getting up to speed specifically on information-fusion applications. In addition, ARL has articulated a goal of developing a laboratory-wide network science research program to address such global issues. However, even the early projects observed during this review cycle before such network sciences programs are fully fleshed out, if continued in relative isolation, may not materially advance ARL's capabilities, especially given the large number of other organizations pursuing similar activities. One suggestion might be to mirror the careful experimental setup and evaluation work done for the prior CISD machine translation work and to focus significant attention on three aspects of the human intelligence problem: understanding what metrics are most valuable to the Army in the field, obtaining or developing realistic but unclassified data sets (such as from gang databases or local police databases), and developing rigorous validation procedures that determine the potential effectiveness both of individual algorithms and of hybrid approaches (as was done in the machine translation arena). Of particular importance, and where perhaps only ARL has the time frames and overall expertise, are issues of scale—what happens as such databases grow to huge sizes and/or are created as collections of separate and localized databases. In a related vein, increasing joint activities with the ISD, NSD, and Mobile Network Modeling Institute may be of value, since there is an increasing understanding that the properties of time-changing networks of all scales, from local sensor networks, to the connections represented by Internet traffic, to the social networks exhibited by both civil and terrorist groups, all obey similar properties, and expertise in one may give significant insight into another.

Also, the potential value to the Mobile Network Modeling Institute of using the Blue Radio as a platform for demonstrating and validating underlying models cannot be underestimated. This is a platform that ARL understands (because ARL designed it) and that was built for an Army application (ground-level sensor networks) that has few if any commercial counterparts (cellular telephones, for example, operate several feet above the ground, with communication with antennas that are quite tall and have different power characteristics from those of the Blue Radio).

The development of the Asymmetric Computing Laboratory should open up significant cross-divisional opportunities. For example, one of the concerns regarding the development of lightweight LIDARs is in the associated computation. Investigating the potential to host such applications on alternative execution platforms seems a natural fit.

Other Areas of Relevance to the Directorate

While the weather modeling efforts within the BED are largely state of the art, challenges exist when trying to move beyond the atmospheric physics to the use of such models for building real applications on top of them. Many of these challenges may be overcome by cross-divisional efforts within ARL. Such collaborations may help, for example, in the use of weather data for the routing of UAVs, for which simple but static algorithms that do not account for weather movements, vehicle dynamics, and formation flying are liable to be inflexible and unverifiable. An understanding of the state of the art in route planning, for both military and commercial aviation, is essential, as is an understanding of how to create more robust, adaptable, and dynamic algorithms. Related projects in decision aids for warfighters that try to use statistics or fuzzy logic to incorporate weather conditions into command decision tools seem to suffer from similar problems of pushing classical algorithms too far.

A paradoxical observation apropos of the highly successful engineering of the Blue Radio, discussed earlier, is that this design effort used state-of-the-art, but largely discrete, components to build a prototype that is clearly acceptable for demonstration purposes but is not at the state of the art in terms of implementation as a single chip that would be needed for a deployable system. There is a real need for the Army to take control of the technology, but the problem is a lack of in-house very large scale integrated chip design experience. Digital chip design can be done relatively easily today (at least for the level of complexity exhibited in the Blue Radio), but analog design, especially for the RF links needed to make a single-chip solution, is much more specialized and probably not something that ARL should invest in at this moment. A collaborative effort, perhaps with a CRADA, may be more appropriate. In any case, it may also be of value for ARL to have enough in-house expertise at least to size such chips roughly and then to project how advances in technology may result in improvements over time in system metrics such as size, power, complexity, and heterogeneous integration (e.g., of complementary metal oxide semiconductor cores and analog technologies through emerging three-dimensional technologies).

In prior ARLTAB reviews, the Board has commented on both the importance of high-performance computing to ARL's and the Army's mission and on the need for sufficient resources to target relevant research with maximal long-term impact. The reorganization of the AC&CSD and a specific focus on high-performance embedded systems should help. There also has been a noticeable improvement in the quality of the research at the AHPCRC. However, significant challenges still exist. Clearly there still is a need for additional research and development resources, specifically for developing a professional staff that is capable of building HPC software products which are efficient and application-specific and for a complementary activity to transition such software into real systems. This resource will become increasingly important as successes are obtained in bringing new supercomputing hardware (e.g., multicore) into the embedded system space.

There is still a lack of an HPC vision in the ARL divisions other than AC&CSD, a lack that will impede the Army's capability of migrating new applications—for example, advanced weather codes or hybrid language translation systems—into combat systems for direct battlefield use. The question is still open on how to leverage systematically both the embedded and the two supercomputer facilities across all activities, including crossovers into nontraditional computing-intensive applications such as signals intelligence. The Mobile Network Modeling Institute is perhaps the first instance in which the use of HPC resources has not just been part of individual projects but has become a unique enabler that is essential to achieving the institute's mission. While serious attempts to use emerging computing devices are laudable, as in the establishment of the Asymmetric Computing Laboratory, there is the danger that more efficient alternative algorithmic approaches may be overlooked. Evidence of this is the recent Stanford vehicle that won the Defense Advanced Research Projects Agency's autonomous vehicle grand challenge by using simple machine learning techniques rather than complex and very computing-intensive, specialized image processing.

Crosscutting Issues of Relevance to the Directorate

Almost all of the crosscutting issues discussed above are of direct relevance to CISD activities or have aspects that could benefit strongly from CISD involvement. The crosscutting issue of microrobotics, for example, has strong roots in all CISD divisions. Clearly, autonomous control, location-identification and trajectory control, and surveillance sensing are squarely in line with current ISD activities. NSD's potential involvement ranges from communication with an individual microrobotic vehicle to managing the behavior of a swarm of such vehicles. BED has involvements on two ends—in using data from microrobots (especially micro-UAVs) to support real-time battlefield weather forecasts or to make predictions about chemical, biological, radiological, and nuclear plumes, and as a user of such forecasts in computing flight paths compatible with vehicle capabilities. AC&CSD is moving toward expertise in embedded high-performance computing—the kinds of computing that would be needed both aboard a microrobot and in the command-and-control links needed to link it into the battlefield. In addition, the ability to do high-fidelity design, simulating, and modeling of the microrobotic platforms during the design phase and in mission planning and rehearsal would be of significant value.

In the next major crosscutting issue, power, CISD clearly must be a key player, both in using information processing to help optimize the power used by a platform overall to perform its mission and in developing information processing systems that are energy-efficient in their own right.

Similarly, in the areas of prognostics and diagnostics CISD needs to be involved both in platform-based fault detection and reconfiguration and in remote real-time data mining, parameter extraction, trend analysis, and real-time modeling.

While CISD does not have as central a role in biomechanics as that of the Human Research and Engineering Directorate, there certainly will be a need to develop and then support significant modeling activities, particularly using HPC expertise, facilities, and resources.

Acoustics is an area that has already been mentioned as a strong point in CISD's research portfolio. This importance will increase as additional sensors and additional laboratories such as HRED's Environment for Auditory Research (EAR) come online and require modeling support, data visualization, and correlation with atmospheric effects.

Modeling and computational sciences clearly overlap multiple components of CISD's charter and have been an area identified for crosscutting activities in previous Board assessments. They remain so.

The issue of identifying potentially disruptive technologies that might radically change the problems confronting the Army (such as the rise of asymmetrical warfare and IEDs) and the way that the Army needs to leverage technology to respond to them are of laboratory-wide importance. However, in the very fast-moving areas that are the realm of CISD, technology changes, representing both threats and opportunities, occur faster than in most other areas. Therefore, each of CISD's divisions, and CISD as a whole, may benefit from an explicit recognition of the potential of such technologies and the development of a formal mechanism to help identify them in a timely fashion.

OVERALL TECHNICAL QUALITY OF THE WORK

One of the assessment criteria applied by the Board asks if the scientific quality of a directorate's research is of comparable technical quality to that executed in leading federal, university, and/or industrial laboratories both nationally and internationally. As in prior years, the answer to this question is that it is generally true for the Computational and Information Sciences Directorate, with exceptional expertise in selected areas such as weather, intelligent optics, and machine translation of foreign languages. Some particularly strong projects reviewed in this cycle were, however, staffed by university scientists, not ARL personnel, and it was unclear how much of the research had been done by or transferred to ARL, and to CISD in particular. Other areas, such as networking sciences, where formal expertise was recognized as needing growth, have seen significant additional resources and organizational changes made to improve them. However, there are still some key areas, such as the capability to develop and deploy HPC-based applications and multicore programming, where additional and improved expertise would be broadly beneficial.

While the CISD's scientific and engineering staff are, on the whole, conducting and publishing quality research in a number of areas, there does not seem to be much involvement in leading scientific societies and organizations or sufficient attendance at top-tier research conferences. Promoting such involvement should give rise to more scientific recognition and stature for the research staff, make them more aware of the state of the art in other groups, and make the laboratory as a whole more attractive to new Ph.D.'s. Sufficient funding should be provided to ARL so that funding is not a constraint on managers' ability to enable the interactions of ARL staff with the scientific community through travel to professional meetings.

A second criterion applied by the Board asks if the research program reflects a broad understanding of the underlying science and research conducted elsewhere. The answer here is mixed; the areas mentioned above as being exceptional are also the areas in which there is a good understanding of the state of the art elsewhere. This is especially true for areas that have emphasized testing and evaluation, such as machine-based language translation. However, in other areas such as route planning, use of field-programmable gate arrays, programming global positioning units, and open-source software packages that do not have a history of prior internal projects or collaborations in that area with others outside ARL, there is a distinct drop-off in an understanding of other work or of the availability of existing program packages.

A related criterion addresses the qualifications of the research team vis-à-vis the research challenges. With just a few exceptions, the match seems to be present. In addition, the aggressive effort to hire new Ph.D.'s and to encourage Ph.D.-level work by current employees is a very positive indication. Evidence of this is a series of talks over the past 2 years by several Ph.D. students.

The next criterion deals with the structure of programs in terms of employing the appropriate mix of theory, computation, and experimentation. The results here are again mixed. In cases where projects take advantage of ARL's outstanding test facilities and weave in a feedback path that validates theory and drives more robust algorithm and system development, the results are usually strong, with obvious opportunities for transition. In other cases, where one or more of these features are lacking or a bit weak, the efforts might be less than optimal. Examples include using simplistic or older algorithms for route planning and polarimetric imaging, and modeling the operation of polarized light sensors in non-ideal conditions and misalignment.

In terms of current and projected equipment, facilities, and human resources, CISD continues to have an appropriate mix to achieve success. This is particularly true of the White Sands, New Mexico, facilities for BED; the intelligent optics laboratories of ISD; the mobile networking laboratories and institute recently established for NSD; and the supercomputing facilities. The development of new facilities such as the Wireless Emulation Laboratory and the Asymmetric Computing Laboratory indicates that ARL is serious about being agile in the face of new technologies. One exception is the need for stronger explicit support for computational scientists and professional HPC staff within AC&CSD for research into high-performance algorithm and program development. The support needed for computational scientists and professional HPC staff should be comparable to the relatively strong support currently provided for infrastructure (e.g., running and managing jobs in machine rooms).

Some of the other directorates, such as the Sensors and Electron Devices Directorate, have leveraged the building of new facilities into an attraction for prospective employees. These new facilities within CISD should offer similar attractiveness and should also be used in that way.

In some divisions with exceptionally strong experimental and field work, such as BED, it may be worth considering whether or not increasing the number of technicians might free more of the research staff time for those research issues of most importance to ARL.

The answer to the question of whether the various research teams are responsive to the Board's recommendations is a resounding yes. There have been identifiable organizational changes, especially in the past 2 years, that seem to be directly focused on alleviating problems on which the Board had commented in the past. The NSD and associated institutes and other initiatives are a premier example, organizing around an end-to-end focus on networking in the large. The reorganization of AC&CSD to address the growing appearance of HPC-like functionality in everyday battlefield computing resources is another example. Further, within the portfolio of research projects there have been significant changes, with drops in areas that the Board had suggested were redundant or behind the state of the art (such as nanoelectronic devices) and the introduction of new projects in areas where the Board suggested that there was significant Army mission-relevant potential (such as embedded HPC, networking problems, and bio-inspired applications). This responsiveness has even shown up in the way that individual divisions, especially BED and ISD, report out their research portfolios at the assessment reviews.

There is, however, still room for improvement, especially in articulating both divisional and overall CISD strategic plans and the rationale behind how the research portfolio is adapted to customer pressures while still maintaining a solid and relevant basic science capability. CISD has shown improvement, especially in BED and ISD, but the improvement is not consistent across divisions and does not articulate as crisply as it could. An emphasis on the core long-term relevant scientific problems and an articulation of short- versus long-term strategic goals would help in continuing to maximize the value of CISD's research portfolio to the Army. A suggested additional metric might relate to how CISD's customers perceive the value of their collaborations, with a related discussion of how expectations and requirements are developed in light of such a metric.

In addition to addressing the assessment criteria, there are several other observations in a variety of areas. First, while there seems to be a significant number of collaborations of various sorts, it is often not clear how those collaborations really interact with ARL programs (versus simply being funded grants), and what part of the results reported from the collaborations are due to ARL versus external researchers and contractors. This information is important when trying to judge the overall level of expertise of the ARL staff. The proliferation of CTAs and ITAs in particular represents collaborations that have not had as much review as other activities have had, and the Board cannot properly judge their overall effect on ARL's portfolio.

Second, judging the understanding of the state of the art would be aided by more explicit discussion in reviews about CISD's view of the state of the art elsewhere and by knowing in what metrics one would see improvements as a reflection of success in ARL projects.

The work at CISD continues to be generally well targeted on Army needs. The machine translation work continues to drive deployments into the field and helps in the processing of newly discovered document troves. BED continues to keep its Army and national science niche in defining and predicting the characteristics of meteorological phenomena that are critically important to fixing the properties of the atmosphere on time and space scales relevant to rural and, of increasing importance, urban battlefield situations. The growth in focus on networking at multiple levels correlates directly with the growth in the network-centric battlefield and the need to integrate disparate information sources in real time to support decision making.

Prior ARLTAB assessments have noted the recognized exceptional contributions of the machine translation work. This continues to be the case.

Judging the contributions of much of the rest of CISD to the broader community remains more difficult. There seems to be a significant variance across the divisions in the number of publications, the quality of the publication forums, and the impact of the work. A variety of indexes are used in academia for such purposes; they include the H-index for references and impact factors for publication venues. Data sources for computing such indexes can be found at Web sites such as those for Googlescholar, ISI Web of Science, Science Citation Index, and Citeseer. Performing such self-evaluations in advance of reviews would help both the Board and ARL to identify where the lead contributions are coming from and which venues should be targeted to maximize the exposure of research results.

3

Human Research and Engineering Directorate

INTRODUCTION

The Soldier Systems Panel of the Army Research Laboratory Technical Assessment Board (ARLTAB) reviewed programs of the Human Research and Engineering Directorate (HRED) within the Army Research Laboratory (ARL) during visits to HRED's primary site at Aberdeen Proving Ground, Maryland, on July 17-19, 2007, and on June 23-25, 2008. A subset of the panel met with several of the HRED scientists to review the quickly developing program in neuroergonomics on both December 17, 2007, and June 9, 2008. In the briefings during 2008, the HRED presenters acknowledged benefiting from earlier comments of the panel, and this was evident in their subsequent presentations.

As general background, HRED is organized as two divisions to conduct research and development efforts to enhance soldier performance. The Soldier Performance Division conducts a broad-based program of soldier-centered basic and applied research and technology testing and evaluation directed toward maximizing battlefield effectiveness. In contrast, the Human Factors Integration Division conducts laboratory and field data analyses, develops modeling and human simulation programs, and performs applied research to ensure that soldier performance requirements are adequately considered in technology development and system design. Tables A.1 and A.2 in Appendix A respectively show the funding and staffing profiles for HRED and the other directorates and indicate the relative levels of effort now devoted to basic and applied research and service activities.

The framework for the assessment, as presented by the HRED Acting Director, emphasized HRED's dual objectives of providing science and technology to enable transitional capabilities for the smaller, smarter, lighter, and faster future force, while also seeking opportunities to accelerate technologies directly into the current force. The HRED Acting Director has been applying a new analytical planning framework, referred to as the Mission and Means Framework (MMF), to plan and coordinate various projects. This approach, although new and unproven, may have merit in that it provides a strong mission-oriented context for various projects; there is concern, however, that it may stifle creative thinking and breakthrough innovations by HRED scientists who wish to study more fundamental problems that do not easily fit into a particular defined mission.

In other words, HRED must more seriously consider ways in which it can best achieve the goal of performing cutting-edge applied research in addition to maintaining its cutting-edge basic research, given that the large majority (approximately 90 percent) of its funding is from 6.2-level or above sources, which target applied research. Because of the existing funding allocation for mission-oriented work, it is important to begin this discussion by revisiting the definition of "applied research," which in the Army is defined from the *DoD Financial Management Regulation* which states:

> Applied research is a systematic study to understand the means to meet a recognized and specific need. It is a systematic expansion and application of knowledge to develop useful materials, devices, and systems or methods. It may be oriented, ultimately, toward the design, development, and improvement of prototypes and new processes to meet general mission area requirements. Applied research may translate promising basic research into solutions for broadly defined military needs, short of system development. It includes studies, investigations, and non-system specific technology efforts. The dominant characteristic is that applied research is directed toward general military needs with a view toward developing and evaluating the feasibility and practicality of proposed solutions and determining their parameters. Applied research precedes system specific technology investigations or development.[1]

A significant proportion of HRED staff resources is allocated to the support of Army programs under development and related field work; that work is not assessed by the Board, whose focus is on the research supportive of the program and field support. HRED is in a unique position to improve the Army's development and use of advanced technologies with targeted, cutting-edge, applied research that will assist in determining the feasibility of various weapons systems and design concepts early in the development process. More specifically, HRED is developing new methods, models, and human performance databases to aid in designing specific Army technologies and in evaluating and improving existing mounted and dismounted soldier systems. To excel in such endeavors, however, will require HRED to continue to perform complex, human-centered, scientifically sound studies that are motivated and defined by a staff and management that have a deep understanding of the complexity of various soldier-system interactions associated with tasks performed within emerging hardware and software technologies.

CHANGES SINCE THE PREVIOUS REVIEW

Five major changes are evident in the Human Research and Engineering Directorate since the previous published review:[2]

1. *Planning for a major new program in neuroergonomics.* This initiative is expected to result in fundamental studies of how to use newer neurological monitoring methods and data along with cognitive and human performance models to improve the decision-making and performance capabilities of soldiers in the field. Because this is such a new area of investigation for the Army, a collaborative program that includes multiple universities, using a Collaborative Technology Alliance (CTA) arrangement, will be the best way to meet the high expectations for this program over the next 5 to 10 years.

---

[1] Department of Defense, *DoD Financial Management Regulation,* Vol. 2, Ch. 5, Washington, D.C., June 2006.

[2] National Research Council, *2005-2006 Assessment of the Army Research Laboratory,* Washington, D.C.: The National Academies Press, 2007.

2. *A continued increase in the number of HRED-relevant projects that have been funded by the Army Research Office (ARO).* In 2004, the ARO funding for such projects was only about $0.5 million. In 2006 it was almost nine times higher, at about $4.5 million, and in 2008 it will have increased again, to about $7.5 million. Such a trend appears to support a strong interest by the scientific community at large in the types of research areas being investigated within HRED. At a minimum, the ARO projects provide access to a broad array of experts for future collaboration. The Board looks forward to learning more about such collaborative efforts and hopes that these result in additional joint (ARL and outside collaborators) publications, visitations, workshops, and seminars.

3. *The near completion of the very sophisticated Environment for Auditory Research (EAR) facility.* This facility is composed of five different sound exposure laboratories and a control room. The facility can present to subjects a large array of auditory experiences, some of which may be unique to this facility, raising the possibility of conducting original basic research that could not be performed anywhere else.

4. *A shift in the activities within the area of network science research.* The 2005-2006 assessment report recognized the potential value from the work being done in social networks and the cognition area, and it encouraged continued activity that leveraged interactions with others within and outside ARL. Since that report, there has been considerable evidence of the latter in the form of grant writing, workshops, and other activities discussed more completely later in this report. A very positive development also has been HRED's creating collaborative research efforts, such as the Multidisciplinary University Research Initiative (MURI) with Central Florida University, establishing the Davis Fellow position, and initiating other joint projects with university faculty and independent researchers in this area. HRED staff have focused their attention on developing a network science portfolio that leverages multiple sources of funding to support basic research on the cognitive and social impacts of networked operations. This includes a basic research (6.1) Network Science Army Science Objective (ASO) that is being funded through ARO and a significant new Network Science Collaborative Technology Alliance funded through the Computational and Information Sciences Directorate (CISD). Notable in these thrusts is the emphasis on multidisciplinary collaboration, as well as outreach to leverage the expertise in academia and industry to advance the state of the art in network science as it applies to emerging needs of the Army.

5. *A change that is not positive: the shift away from an emphasis on the development of human modeling capabilities that were highlighted in the 2005-2006 assessment report.* There have been some enhancements in the usability of the Improved Performance Research Integration Tool (IMPRINT), which is the primary human-performance-modeling algorithm and software developed by HRED to predict soldier task time requirements and mental workloads. These enhancements include improved IMPRINT usability (e.g., improved formats of outputs and better accessibility) and functionality (e.g., enhancements to workload scales) over the past 2 years, during which congressional funding supporting the IMPRINT program shrank (from approximately $2 million in FY 2006 to $1.14 million in FY 2007) and then evaporated (replaced in FY 2008 by $270,000 in ARL mission funding). The FY 2008 funds have been devoted largely to the maintenance and upkeep of IMPRINT, and there appears to be a risk that HRED will continue its recent practice of largely restricting its use of the IMPRINT human task analysis model in its current state to perform first-cut human factors analyses of anticipated hardware design problems. Although the IMPRINT model is useful in its current state, many different empirical studies are underway or have been completed within HRED and elsewhere that could continue to enrich and improve the IMPRINT model, and this was encouraged in the previous ARLTAB assessment report. Unfortunately, little has been done to translate new empirical findings into the model to make it more robust and valid for future design analyses. Furthermore, it was not clear that the existing model is being used very often to structure and plan future empirical studies conducted within HRED. An exception

to this was a study conducted in 2007 of the task demands on a gunner in a mounted combat system. IMPRINT was used to determine whether this soldier could also control an unmanned ground vehicle (UGV) while moving. The IMPRINT model showed that it might be feasible, and in so doing provided several well-structured hypotheses for subsequent study. Follow-up empirical studies indicated that the gunner would miss too many targets while attempting to control the UGV (see the later discussion of human-robot interactions). This example is highlighted here because it illustrates two common research issues: the model was helpful in the early planning of a complex empirical study, and the IMPRINT model's ability to make accurate predictions about human behaviors needs to be improved in order to better simulate certain types of complex soldier tasks.

ACCOMPLISHMENTS AND ADVANCEMENTS

Environment for Auditory Research Facility

The completion of the Environment for Auditory Research facility is a very significant event. The listening laboratories in this facility will allow unique studies of how different sounds are identified, localized, and used for communication in a variety of environments. Given the uniqueness of the five different physical listening laboratories, however, it is very important to specify the physical acoustics of each of the rooms and publish the results soon so that other experts from a variety of disciplines can appreciate (and possibly use) the facilities in the future. In terms of collaboration, the EAR facility group is in conversation with one or two research groups, including the Air Force acoustics research group. Work within the Sensors and Electron Devices Directorate (SEDD) on acoustic devices and algorithms may also be applicable. This type of cooperation is certainly needed in such an important area.

One of the advances that this facility provides is the opportunity to examine auditory capabilities and communication limitations in a variety of very complex acoustic environments, both real and virtual. However, research in auditory communication performance may not exploit the full capabilities of this facility. From a medical standpoint, loss of hearing is a major issue in the Army. In this context, it is good to know that this facility has a medical audiologist on the staff, particularly given the strong interaction between the ability to segregate sound sources and hearing loss. However, collaboration with other groups concerned about noise-induced hearing loss should be given some priority. Groups at the U.S. Army Aeromedical Research Laboratory at Fort Rucker, Alabama, are performing studies to document the full extent of health hazards due to noise exposure in the Army and are developing programs in accordance with Army Regulations 40-10 to provide systems and materials to preserve hearing. Integrated hearing protection and hearing communication systems also are being developed at Fort Belvoir, Virginia, in the Program Executive Office-Soldier Programs. All of these would appear to be relevant in the planning of future research in this new facility.

The EAR facility staff is quite strong, but it is not clear how the staff will balance applied development work and publications in peer-reviewed journals. In the past they have published a good number of technical reports and conference proceedings, with some on the Internet, but more publications in peer-reviewed journals are necessary in order to gain the credibility and attract and retain the best people to this new facility.

As for the physical facility, of the three anechoic rooms the Distance Hall is the most novel anechoic chamber (for human psychophysical testing) by virtue of its size and configuration. The multiple speakers allow the simulation of variously placed sound sources and reflective walls. As a result, this space provides an opportunity to examine the psychoacoustics of distance perception and motion perception, as well as to answer more fundamental questions concerning the psychoacoustics properties of environments with multiple (virtual) reflections and sources. This space provides a unique opportunity to evaluate the sensitivity of human observers to differences in sound source depth, as well as changes in depth (motion). The speaker (distant) arrays also provide virtual walls to study complex sound reflection effects on depth perception.

The other two anechoic spaces are impressive in size and development. The Sphere Room appears to be slated for basic and applied sound source localization research, including the rapid measurement of head transfer functions both with and without helmets. The Dome Room has characteristics appropriate for experiments concerning the perception of sounds in space and the motion of sounds in azimuth with sound reflections.

The new physical facility is impressive, and several well-qualified staff members are in place. It is not as clear, however, that they have identified the research questions that this HRED facility can uniquely answer. For example, What are the classic questions in psychoacoustics that can be revisited in this new facility? How will the staff see the work through from conception to publication? For each component of the laboratory, the general tenor of the research to be conducted was provided to the Board, but details were lacking. This was particularly true of the Listening Laboratory, where the description was limited to the auditory capabilities of the space and not the science that it will subserve. The overwhelming weakness was the absence of a coherent research plan as to how the EAR facility would attain its lofty goals. Given that construction has been in the works for several years, one would expect a complete description of the research that is to be initiated over the next couple of years, including the hypotheses to be tested, the measurements to be made, the source of experimental subjects, and other experimental factors. The absence of details at this time is a concern and in a few instances was alarming. In summary, EAR is a wonderful facility, and some excellent people are in place. Now is the time to reach out to other experts, to set up workshops and visitations, and to plan carefully an equally strong program of both basic and applied research that will result in this being seen as a first-class national resource in the future. HRED should consider collaborating with acoustics experts among the staff of the ARL Sensors and Electron Devices Directorate.

Night-Vision Research

Night-vision enhancement technologies are rapidly being deployed in both the civilian and the defense sectors. HRED has been involved in human factors studies of night-vision devices for many years, but perhaps because of the limited nature of the briefings provided on this topic, it is not clear how HRED's work fits in with the broader work in this area. A central concern has been sensor fusion—the need to combine multiple sources of information into a display that a user can interpret. Groups in many fields are working on these issues. Military laboratories in several countries publish in this area. In addition, major automobile-manufacturing groups are involved in fundamental studies of night-vision enhancement technologies. It is not clear whether the HRED researchers have collaborated or consulted with the many groups around the world that are studying this problem. Certainly collaboration would be appropriate with the visual scientists and engineers at the Army's Night Vision Laboratory at Fort Belvoir. An HRED-organized workshop on this topic might provide a means to focus some of the activities better.

The research on sensory fusion that was reported, while of importance, does not seem likely to make significant contributions to an understanding of the perceptual issues in sensor fusion. The specific projects do not appear likely to produce results that could be published in the peer-reviewed literature—an indication that the approach being taken is not current. For example, the sensor fusion algorithms under study were very basic. A quick survey of the literature suggests that other laboratories are experimenting with much more sophisticated algorithms for cue combination. Many of these seem to be based on Bayesian approaches in which an effort is made to determine which of several signals is providing the most information at *this* location and at *this* time. Similar approaches are used in other areas of vision research (e.g., models of eye movements and models of depth cue combination). Once again, contact with researchers who work on cue combination, Bayesian theories, and ideal observer approaches would be useful. In this context, some of this collaboration could be established by greater participation at national vision and/or machine vision conferences.

The Board, in referring to past HRED research, especially that discussed in the 2005-2006 assessment report, noted that when presenting an image to one eye that is different from the image visualized by the other eye, a phenomenon known as sensory rivalry occurs. This can result in perceptual errors and cognitive distraction, particularly when dynamic displays are involved. The research reported during the current assessment did not deal with this important topic specifically, but rather relied on eye-movement studies to indicate how and when sensory fusion would occur. It was not clear what hypotheses were being tested with this approach. The work being done is not being published well. Although HRED has produced several conference proceedings and technical reports, there are only a few papers in refereed journals over the past several years. The dearth of significant publications should be of concern both to the responsible researchers and to those responsible for directing the overall effort.

Cognition and Neuroergonomics Collaborative Technology Alliance

The panel's select group of experts on neuroergonomics was very favorably impressed with the current accomplishments of the neuroscience group in HRED. The use of the CTA mechanism, joining industry and academic groups, is appropriate. The HRED group now includes outstanding new, young scientists with excellent backgrounds. The motivation of the group appears high, and the members have demonstrated major advances over the past year in their understanding of the field and what they can contribute.

The neuroscience group's organization of the 1-day workshop held on May 8, 2008, was an excellent method of informing neuroscientists and cognitive scientists of the Army's needs and of quickly evaluating various research groups that could respond to a CTA announcement. This workshop allowed 18 different research groups to present projects (self-selected) on a variety of topics relating to applications in the broad field of neuroscience to address Army needs. Some were excellent—for example, the project on independent component analysis of electroencephalography. Others were less so. The HRED scientists reviewed these presentations at a special meeting on June 9, 2008, with a subset of the panel members. The HRED staff indicated that the pending CTA announcement would not cover all areas addressed by the workshop attendees but would focus on a few areas that seemed most promising for basic research and that would have clear applications to the Army's needs. In particular, the staff indicated their intention to emphasize the development of a portable monitoring system of neurological functions that could predict when a soldier's cognitive abilities were being overly stressed by particular tasks and the environment to the extent that performance would be compromised. The details of the CTA announcement were not available at the time that this report was drafted, but HRED should develop advanced cognitive performance models to form the basis for hypotheses to be tested, and to relate various neurological measurements to the prediction of human performance capabilities and mental workload.

Human-Robotics Interaction

Current military operations are employing robotic systems in unprecedented numbers and roles. Many types of unmanned aerial vehicles (UAVs) are used for intelligence, surveillance, and reconnaissance (ISR); for targeting and for tactical intelligence; and even for tactical engagement. Unmanned ground vehicles are being employed to deal with threats. Most of the systems in use require the attention of at least one or more operators for each vehicle. The potential ground combat leverage available through unmanned and robotic systems will never be realized until the ratio of operators to unmanned systems is reduced. Although the use of multiple robotic systems that incorporate all types of unmanned vehicles in military operations has been advocated, multiple robotic systems are unrealistic under current operating protocols. The Army's Future Combat Systems include, conceptually at least, multiple types of robotic components; their leverage could be increased significantly by one-to-many (human-to-robot) control. Achieving technological superiority through the one-to-many control paradigm is one key approach to countering the higher leverage obtained by potential enemies employing an asymmetric approach to engaging U.S. forces.

During the 2007 meeting at HRED, a number of HRED human-robotic interaction projects were described. The HRED human-robotics work was of relatively high quality. Much of the presented research focused on single platform systems, however. The research focus was on fundamental human-robot interaction questions, unlike the previous review during which much of the research focused on robotic perception, mission packages, and supervisory control of single platforms. The HRED group has a unique opportunity to begin a program of research that focuses on understanding how multiple UGVs per operator can be controlled in real-world scenarios (see the discussion in the "Opportunities and Challenges" section below).

Network Science: Social Networks and Cognition

As the 2005 report of the National Research Council entitled *Network Science* points out, interacting networks in the physical, information, cognitive, and social domains are ubiquitous in U.S. military operations, and they are increasing in importance with the growing efforts to transform the U.S. military into a force capable of network-centric operations.[3] In this context, the domain of network science is defined broadly within ARL. It encompasses the range of phenomena emerging from the introduction of network-centric operations and needing to be understood and addressed to support the warfighter. Such phenomena range from understanding the physical characteristics of networked operations (e.g., characteristics of sensors and radios), through the communication level (routing, self-configuring networks), through the information level (e.g., secure information flows), and on up to the impact of network characteristics on the performance of individuals and teams operating in a networked environment. HRED research activities and goals in network science cover social network research as well as research on the impact of network operations on human cognition. The early in-house studies in network science presented by HRED in 2007 were not very impressive, but they did allow the HRED staff to become more familiar with the methods and models needed to perform high-quality research in network sciences. It would appear that, appropriately, the staff has now planned a much more aggressive program of research, as noted below.

HRED is to be particularly commended for its success in leveraging ARO funding to advance the basic research foundation relating to human decision making as it is affected by Army command-and-control structures. An ASO program under ARO funding currently planned for the 2008-2012 period

[3] National Research Council, *Network Science,* Washington, D.C.: The National Academies Press, 2005.

will involve conducting experiments and observations to explore interacting network effects on human decision making.

HRED is successfully leveraging ARO funding to foster greater interdisciplinary dialogue within and outside ARL. For example, a Davies Fellowship under ARO sponsorship has been obtained to fund a mathematics professor from West Point Military Academy to work with HRED behavioral scientists to explore the impact of network structures on human decision making. HRED also organized a Network Strategic Technology Initiative workshop that brought together leading researchers in network science to contribute to the Army's understanding of the state-of-the-art research in network science and its relevance to warfighter issues.

HRED, in conjunction with CISD and SEDD, is initiating a major new Network Science Collaborative Technology Alliance that is intended to develop the basic research foundation to enable modeling, design, analysis, prediction, and control of secure tactical communications, sensing, and command-and-control (decision making) networks. This includes research on the impact of networked processes on individual and distributed team decision making, from both individual cognitive and social network perspectives. The research thrust to be taken is in the planning stages. Nevertheless, this is clearly an important research initiative that has the potential to advance the state of the art in network science as it relates to critical Army needs, and the Board looks forward to following the progress of the research.

HRED is to be commended for initiating 6.2-level network science research that is directly focused on an urgent Army research need—analyzing the impact of complex dynamic network-centric environments on individual and team cognitive and collaborative performance. As explained in HRED's description of its 6.2 Tactical Human Integration of Networked Knowledge program, network-centric operation involves an abundance of information received from many different sources (human and sensor) and presented across multiple modes (text, audio, visual). Human decision makers are currently unable to make effective use of this information for reasons including information overload as well as network bandwidth and other hardware constraints that introduce lags, information loss, and degradation, leading to performance problems that include attention misdirection and poorly calibrated trust in the information received. HRED is initiating a multiyear program—the Tactical Human Integration with Networked Knowledge (THINK) Army Technology Objective—scheduled to start in FY 2009 and to extend through FY 2012, to analyze the contributors to performance problems and identify methods for overcoming these problems. This work will bring together experts in cognitive science, social network science, and computer science to work collaboratively to develop and evaluate methods to train and improve information sharing, decision making, and collaboration in networked operations. It will also aim to develop improved methods and guidelines for information aggregation and alerting so as to enable more effective attentional focus on high-priority issues, produce better trust calibration, and improve the quality of individual and distributed collaborative decision making. This program includes laboratory and field experiments intended to validate the impact of proposed enhancements in training, social network organization, and new information-aggregation and alerting concepts.

Biomechanical Modeling Research

In the previous ARLTAB assessment report, it was noted that biomechanical analysis tools were used to understand the basis for injuries incurred by persons lifting heavy components during bridge building. The ability to merge new cognitive models and biomechanical models of soldiers into the IMPRINT framework has been recommended in the past, although apparently funding has not been provided recently to pursue this development. Nonetheless, two recent biomechanical studies both deal with very real and important problems for the Army and use biomechanical modeling methods that could in the future enhance the physical-task-analysis models within the IMPRINT program. The first of these studies explored the effects of the mass of various handheld weapons during dynamic targeting motions. The study had not been completed, but the biomechanical analysis methods were appropriate, and preliminary results showed that adding mass to today's weapons has a deleterious effect on quick aiming motions. The second biomechanics study was to determine the injury risk factors during long marches. It explored how certain foot and leg motions could produce stress fractures in the lower leg bone. Once again, biomechanical models were used to understand subtle motions and provided a means to predict how certain types of gait motions raised the risk of injury. These continuing studies illustrate the use of human modeling methods to plan the research, analyze the results, and eventually provide simulations of soldier performance characteristics in a variety of task scenarios. As in the cognitive modeling area, this type of modeling and empirical research should be continued so as to produce more useful and accurate simulations of soldiers' physical endeavors.

OPPORTUNITIES AND CHALLENGES

It should be clear from the preceding sections of this chapter that many worthwhile applied research and development actions have taken place since the previous review. There are many opportunities and challenges that exist within each of these areas. The most significant of these are discussed below.

Neuroergonomics

The opportunity to work in the rapidly developing area of neuroergonomics is noteworthy. To lead this effort, HRED has assembled an excellent team of researchers who have engaged many others in the field to define a coherent program of basic and applied research. This approach is appropriate. The challenge will be to find the effective mix of neurological, cognitive, and human performance scientists and to bring them together in a collaborative team to further the understanding and modeling of human decision making and actions that affect a soldier's effectiveness. It also will be necessary to ensure that the HRED-supported research is complementary to a number of other similar studies supported by the Department of Defense in this arena and to ensure that this complementary research is supported by ARO. Advanced cognitive performance models should be developed to form the basis for hypotheses to be tested and to relate various neurological measurements to the prediction of human performance capabilities and mental workload.

Environment for Auditory Research Facility

The near completion of the EAR facility provides another opportunity for HRED. This facility could allow the researchers using it to become leaders in the area of auditory performance. The team of researchers appears to be well prepared to perform cutting-edge auditory studies. Their challenge will be to develop a formalized and coherent set of studies that take full advantage of this outstanding facility, while meeting the unique needs of the Army both to protect soldiers against noise-induced hearing loss and to improve auditory communication and performance. There is a risk that both investigator and laboratory time could be absorbed by short-term practical questions, such as how different helmet designs influence sound localization. If time is not set aside for more exploratory basic studies, the scientists may not remain at the cutting edge of the research, and it will be difficult to attract the best scientists to the laboratory, thus losing the advantage now provided by such a well-conceived physical facility.

Vision

During the 2007 briefings, HRED discussed the problem of binocular rivalry. This phenomenon arises when different images are presented to a person's two eyes. This is a specific problem within the broader problem of sensor fusion and cue combination. It arises in HRED settings when vision enhancement and communication devices present one set of imagery to one eye and another set to the other eye. This is a very important problem, and HRED has many opportunities to enhance the knowledge base in this area. Given that this was a research topic in the past, it is disappointing that no results seem to have appeared in the scientific literature. More generally, the rate of publication in this area is very modest. The list of publications during the 2005-2008 period appears to show 10 publications from the vision group. Of these, only one has been submitted to a peer-reviewed journal. All of the other papers appear to consist of conference proceedings or ARL technical reports. In general, the vision group is working on interesting problems that have very significant basic and applied potential. The information presented by HRED suggests that this potential has not been realized. If the group is to continue working in this area, it should develop much more extensive contact with the broader community working in this and related areas, and HRED should consider the sort of commitment of resources and personnel that have positioned the auditory group to make significant contributions in this arena.

Network Science

In the network science area, major opportunities exist at ARL generally and in HRED in particular to build on their existing unique network capabilities. More specifically, the opportunity for ARL is to create a domain that captures the unique characteristics of the research laboratory by assembling staff that can address the cross-disciplinary problems inherent in this area and ensuring that they address issues affecting the mission of supporting the soldier. Network science methodology in general and network metrics in particular are still in the early stages of development. ARL is well positioned to advance the state of the art in these domains if it can pull together an effective team of people from the physical, mathematical, software, and social and behavioral sciences. Unlike many other research institutions, ARL can gain access to network data gathered from varied simulations and field exercises as well as from real interactions among network members—these members including not only soldiers, but also robots and other network-based information agents. Essentially, such contexts present unparalleled opportunities to develop new network research paradigms as well as to assess the reliability and validity of existing network metrics. The primary challenge for the network science group at HRED is to execute the current research projects and become more involved with future streams of research. A great deal of energy is being expended and much activity is occurring, but there is little completed work in this area available for evaluation by the Board. This is understandable, because the network science program is early in its development. A second challenge is to move interesting, ongoing efforts and future planned work to completed outcomes, and to publish this work quickly so as to establish and enhance credibility with peers in this arena.

Workload Modeling

The HRED human factors design aid referred to as the IMPRINT workload simulation model continues to represent a major success story for the directorate. It appears not to have evolved in functionality a great deal since the previous ARLTAB review. This is unfortunate, because there is a great need to have a robust human factors (HF) analysis tool for planning research studies and for assisting Army contractors in meeting HF requirements within the MANPRINT (Manpower and Personnel

Integration) program. The challenge will be to prioritize the functionality most needed and to develop software projects that would most effectively meet the Army's needs to improve the design of future human-hardware task systems and to better train future warriors. To do so will require that the IMPRINT model be better able to simulate the effects of various complex perceptual and cognitive tasks, perform dynamic biomechanical motion and vibration simulations, and provide predictions of operator effectiveness and mental loads when the operator is controlling multiple UGVs. There also is the need to develop an IMPRINT model that is capable of analyzing the workload imposed on soldiers while operating as a team, and for long periods under high mental and physical loading. There also is a need to understand what type of formal training is required so that users can fully implement and accurately use the IMPRINT model to simulate various design scenarios, especially since the model can be operated with different levels of complexity and functionality.

Robotics

As has been stated previously, a real benefit of using UGVs will be realized when several UGVs can be controlled by a single operator. Although the present work on the interactions of an operator and a single UGV are extremely important, ARL should consider developing an enterprise-wide program dealing with the semiautonomous coordination of multiple robotic systems (i.e., the networking of information and autonomous actions taken by the robotic systems with varying degrees of human intervention to further mission objectives). To accomplish such coordination will require research to further define the supervisory control structure for groups of UGVs and UAVs, using smaller groups of human operators than are now required. Building on a strong foundation of research on individual robots, ARL also can bring to bear significant talent from HRED to address human-system integration, along with robotics-related work in the Sensors and Electron Devices Directorate and the Computational and Information Sciences Directorate. This cross-directorate work will provide a means to address actual, rather than simulation-based, platform-to-platform and platform-to-human communications and mutual awareness.

A key advantage of this approach is that it will position ARL to serve Army needs in a large variety of contexts and concepts of operations, regardless of the specific properties of the individual unmanned systems that are ultimately developed or acquired by the Army. It is believed that commercial developers are outpacing the Army's in-house efforts in individual robotic systems, despite the fact that multiplatform coordination and supervisory control of large numbers of robotic systems represent new frontiers that for now are relevant mostly to military rather than commercial users. Strategic application of resources to these problems can position ARL to enable the Army to excel in the operational use of unmanned, remote robotic systems.

Biomechanics

As to research addressing the physical requirements of manual tasks of soldiers, there is a very large opportunity to enhance the modeling of the musculoskeletal system of soldiers in order to predict their physical performance capabilities in such tasks. Most of the existing modeling has relied on overly simplistic, structural representations of both the anatomy and the physiology that govern human exertions of all kinds. The types of complex perceptual-motor tasks required of soldiers demand that the highest quality of biomechanical modeling and empirical studies be available if the Army is to understand the environmental and task factors that affect a soldier's performance capabilities. The staff at HRED appears to have the fundamental biomechanical knowledge and some of the physical resources necessary to push forward on such research. The challenge will be to structure the research in such a way that it provides further insights as to how the physical models now used as part of the IMPRINT program can be improved. Because a number of academic and industrial research groups are working in the area of biomechanical modeling, a workshop should be held to explore further which types of existing biomechanical models are most appropriate to guide and enhance the types of research most needed by the Army.

OVERALL TECHNICAL QUALITY OF THE WORK

There is much to admire about the progress that has been made over the past 2 years at HRED. In particular, there appear to be some new, well-trained researchers on the staff who understand the need to perform applied research that is of high quality in a scientific sense. In this context, the near completion of the EAR facility and the new funding being provided for the neuroergonomics program provide new resources (physical and financial) that are unique. Unfortunately, since formal research plans were lacking in detail, it was made clear to the Board how the staff will best use the new resources in the EAR facility and the new funds for the neuroergonomics program to balance research that could provide both scientific breakthroughs and solve important military problems.

If one looks at the publications coming from HRED over the past couple of years in all the areas, not just the two areas mentioned above, the number of peer-reviewed journal papers is not very impressive. Most of the publications are in technical reports and proceedings. These are certainly helpful: they show that the staff is capable of reporting their methods and findings to a limited extent, and they can represent the only viable outlet for some of the directorate's research that is not experimental or that involves single large exercises or simulation experiments. However, such reports often lack the clarity needed for others to fully evaluate and hopefully come to respect the cutting-edge work being done in HRED.

The six areas of concentrated research reviewed in this chapter are of vital importance not just to the Army, but to society in general. This list includes the following: (1) providing a better understanding and the means to enhance audiometric performance—a major problem for older individuals; (2) understanding neurophysiology at a level that predicts when a person is cognitively incapable of performing certain tasks; (3) presenting networked information to people in a fashion that can be quickly and accurately understood and acted on by one or many people; (4) being able to control multiple unmanned vehicles and tactical resources with minimum human interventions; (5) being able to understand and predict the physical capabilities of soldiers to perform complex and fatiguing manual tasks; and (6) providing the means to be able to perceive objects while in darkened environments. These all are very important and scientifically challenging problems. They all require study by teams with multidisciplinary backgrounds, which appear to be available in HRED for most of the areas. However, closer working collaborations (e.g., co-authored papers with experts outside HRED) are needed in all the areas to complement and enhance the capabilities of the staff. In some cases this is being done well, but not in all the areas. As noted at the beginning of this chapter, the closer alignment of ARO and HRED research will address this recommendation.

From a methodological perspective, most of the areas of research conducted by HRED are strictly empirical. In some cases models are used to justify particular types of empirical studies. For instance, the IMPRINT task-analysis model has been used to provide some limited performance and mental-loading predictions associated with the performance of various complex tasks that were being considered for future empirical study. Select biomechanical and workspace-analysis models also have been used to understand the cause of certain types of injuries. HRED should continue this trend; many more future studies should include the use of analytical models during the planning of experiments. Such efforts can often lead to more efficient laboratory studies and, more importantly, can assist in allowing the results obtained from small empirical studies to be compared to other studies to gain general validity and applicability.

Since the origins during World War II of organized research that was meant to understand and model human-hardware system interactions and consequences, the military has been the largest benefactor and supporter of such work. One might surmise that after 65 years of such work, there is not much that has not been addressed in this area. Yet the operational complexity of current military systems, not to mention future systems, demands that one know much more about the mental and physical attributes of the soldiers who are expected to operate and maintain these systems under the most arduous conditions imaginable. George Fisher, former chair of the National Academy of Engineering (NAE), noted in his 2000 address to the NAE, that we are in the Dark Ages when it comes to designing systems that are convenient for people to operate. Indeed, it is estimated by some that fewer than 10 percent of currently graduating engineers receiving a bachelor's degree have had even one ergonomics course, and fewer than about 2 percent of engineers receiving a Ph.D. degree have had such exposure. Given this situation, is it any wonder that the military continues to be plagued by hardware and software systems that are extremely difficult to operate effectively and safely and to maintain?

HRED identified six areas of concentrated research and development and requested and supported this review of those areas. The areas selected are, in general, highly appropriate and important not just for improving military operational effectiveness but also for improving the quality of life for all people. Although this report raises questions about the quality of the research in some of these areas, it is clear that most of the staff are capable of performing outstanding applied research in the various areas reviewed. It also has been acknowledged that the facilities are being improved to support the empirical studies that are needed. Continuing to pursue opportunities for more collaborative interdisciplinary research would contribute further to HRED's studying and modeling of complex real-world conditions. For example, bringing together investigators focused on auditory processing with cognitive neuroscientists might strengthen this area and provide more useful results. A growing awareness of the importance of engaging other similar research groups seems to be taking place in some of the topical areas, but there should be more workshops and visiting senior scientist positions, along with support to publish more papers in peer-reviewed journals and with co-authors from different laboratories.

Over the past 2 years, not much tangible in terms of new major findings has resulted. However, there exists a great deal of excellent potential for HRED to become a first-in-class research organization in several areas. To do so will take leadership that understands and respects the complex issues involved in performing cutting-edge, human-centric research and model development. Such leadership must manage the sometimes conflicting research goals resulting from the need for fast evaluations of new technologies and systems that are being rapidly deployed, versus providing scientifically valid, predictive models, methods, and principles to improve the design of future combat systems. With such leadership, HRED can become an outstanding national resource, given the excellent staff and physical facilities that are beginning to be available.

4

Sensors and Electron Devices Directorate

INTRODUCTION

The Panel on Sensors and Electron Devices of the Army Research Laboratory Technical Assessment Board (ARLTAB) met to review the Sensors and Electron Devices Directorate (SEDD) at the Army Research Laboratory (ARL) facilities at Adelphi, Maryland, on July 18-20, 2007, and May 28-30, 2008. SEDD contains four divisions, all of which were reviewed by the panel: Electro-Optics and Photonics, Radio Frequency and Electronics, Signal and Image Processing, and Directed Energy and Power Generation. SEDD is responsible for the Micro Autonomous Systems and Technology Collaborative Technology Alliance (CTA), which was awarded in February 2008. The Computational and Information Sciences Directorate (CISD), the Vehicle Technology Directorate, and the Weapons and Materials Research Directorate also contribute to the management and to the collaborative research conducted in the CTA. SEDD also has responsibility for the sensor information processing research area within CISD's Network and Information Sciences International Technology Alliance with the United Kingdom that began in 2006.

CHANGES SINCE THE PREVIOUS REVIEW

There is a broad scope of activities within SEDD, encompassing power sources and electronics; acoustic, magnetic, and electric sensors; advanced radio-frequency (RF) technologies; signal and image processing; and sensor fusion. The breadth and scope of projects in the directorate are appropriate to the mission of SEDD and responsive to the Army's needs.

There is a good balance between the pressure of near-term deployable technology development and long-term basic research, and there is a clear awareness of this balance among the management and staff of SEDD. SEDD's discussions of the roles and expected outcomes for the projects, especially those identified as purely technology development, were particularly impressive. The research environment in SEDD is very positive. There is a level of energy and interest that clearly reflects a positive culture and a strong sense of value in the work that is being done.

A healthy, confident culture exists in research activities within SEDD. It appears that a research activity first determines the objective or particular ARL-critical need. Next the ARL internal strength is assessed to determine what resources need to be aligned to conduct the research; identified weaknesses are then strengthened if possible. If the necessary expertise or resources cannot be established within ARL, all technologies that are available and useful are examined around the world. If a desired technology or capability that is external to ARL is identified, collaboration is sought out among established academic researchers and/or industrial entities. And, finally, if a particular recognized research void still exists, the SEDD staff work with the Army Research Office (ARO) to define appropriate research programs, to create and solicit research proposals, and ultimately to fund research endeavors to fill the needed ARL-critical objectives. This culture was evident in quite a number of research activities and is crucial for rapid ARL mission-critical advancement.

Many projects have impressive research in materials, processing, devices, and characterization. To continue to achieve the various goals requires a great deal of infrastructure with capital equipment renewal, as well as expansion of equipment capability. Of course some activities can be carried out using resources external to ARL or within collaborative research activities. In some cases, however, capability must reside in-house. The recent acquisition of a hydrofluoric acid (HF) vapor etching tool for microelectromechanical systems (MEMS) research is commendable. Such a capability will significantly advance and enhance all aspects of MEMS research from the point of view of time to completion for the fabrication of a device, fabrication yield, and the ability to create devices yet to be conceived. Acknowledging the absence of a dedicated equipment budget, the SEDD management team clearly tries to support the equipment needs of the researchers, even in times of reduced overall budgets.

SEDD evaluates its programs and modifies its focus from year to year as necessary to meet the Army's needs. Thus there have been various changes to the directorate's programs over the past 2 years since the previous ARLTAB report[1] as the directorate's efforts were refocused. SEDD has initiated new programs in microsystems, radar biometrics, situational awareness, compact radar, and power sources for unattended ground sensors. In parallel SEDD has intensified its focus on solid-state lasers, vision protection, sensor fusion, flexible displays, bio-inspired materials, antennas, and reserve batteries. SEDD has decreased its investments in magnetics, power MEMS, liquid reserve batteries, and platform RF sensors.

The folding of ARO into ARL appears to be going smoothly. There is a very good connection between the ARO and ARL missions. ARO provides an important liaison role between ARL in-house research and external university research. Universities are made aware of the immediate needs of the Army, and in turn ARL has a natural pathway to make use of the university research results. Undoubtedly there are organization-level efficiencies as well. The synergistic connection of SEDD and ARO was described and is clearly extremely important. Clear communication channels exist between program managers at ARO and all levels of personnel at SEDD. Research needs of ARL activities, once identified and defined, are articulated to ARO to establish research programs with opportunities for contribution by external entities, both academic and industrial. Furthermore, SEDD scientists are welcome to participate in certain ARO-funded programs if desired. Students are funded through ARO fellowships to have internships within ARL. Some ARO program activities, such as the Strategic Technology Initiatives, are chaired by a program manager from ARO in conjunction with an ARL scientist. Activities are

[1] National Research Council, *2005-2006 Assessment of the Army Research Laboratory,* Washington, D.C.: The National Academies Press, 2007.

conducted by ARO program managers and others across the many government services to determine types of programs needed, as well as to define research thrusts. Clearly the close interaction of ARO with SEDD enables SEDD funds and research to be heavily leveraged.

ACCOMPLISHMENTS AND ADVANCEMENTS

Electro-Optics and Photonics

SEDD's work on electrooptic sensors has made significant progress over the past 2 years in several important areas. The infrared detector program continues to carry out excellent research and has demonstrated year after year very impressive accomplishments. The overall goal is to demonstrate advanced cooled and uncooled infrared (IR) detectors and detector arrays for the Army, exploiting a fundamental understanding of the physics and chemistry of various semiconductor compounds.

The breadth of ARL work on materials and devices for IR detection is comprehensive. This is perhaps the only laboratory in the world with the capability and collaborations to realize devices with operations that span the wavelengths across the entire infrared spectrum. The work presented covered various materials systems and devices including the following: II-VI materials HgCdTe IR focal plane arrays on silicon (Si) and high-operating-temperature long-wavelength infrared (LWIR) HgCdTe detectors; III-V materials such as AlGaAs/GaAs quantum-well infrared photodetectors (QWIPs); Type II GaSb/InAs detectors and dilute nitride GaInSbN detectors; and IV-VI materials such as PbSnSeTe detectors.

The breadth and quality of the work reported are impressive. ARL is clearly a leader for infrared detector technology. The work on HgCdTe on Si, corrugated quantum-well infrared photodetectors (C-QWIPs), GaInSbN, and PbSnSeTe on Si are setting the trends. On the C-QWIPs, the image demonstrated with the detector array is striking with a temperature resolution less than 0.022 K. The work demonstrated clearly is moving to satisfy Army needs. SEDD has published two journal papers and three conference papers. Considering the impressiveness of the work presented at the review, more papers should be expected on the work.

The use of dilute nitrides for very long wavelength infrared (VLWIR) detectors is a novel approach for extending the band gap of III-antimonide semiconductors into the LWIR and VLWIR regimes by substituting nitrogen for (a small fraction of) the antimony anions. It is a relatively new high-risk/high-payoff exploratory effort that has the potential to yield important materials results. The probability of deriving fundamental insights from this work could be enhanced through strengthened theoretical support in the areas of electronic structure, growth kinetics, and disorder. A new emphasis was described that takes advantage of the antimony material system alloyed with a small percentage of nitrogen. The material system is completely unexplored and represents a first effort conducted with ARL resources; the ideas and first results are very encouraging. The work in dilute nitrides is new but very promising. GaSb-based materials such as GaInSbN with 1 to 4 percent GaN have the potential for providing a new infrared material system. The idea of integrating electronics and detectors is a good approach based on excellent materials work that continues to improve over time.

The MgCdTe and HgCdTe on Si work is doing very well, with impressive results. The identifying factor of ARL work is the pursuit of the growth of HgCdTe on Si for lower-cost detector applications. SEDD has achieved up to 1 K × 1 K LWIR HgCdTe on Si. SEDD has demonstrated novel C-QWIP devices that have high quantum efficiencies (QE) of approximately 40 percent, which is the highest in the world. The work on Type II GaSb/InAs superlattice is progressing, with challenges on defect issues and passivation technologies. Notwithstanding these issues, devices with approximately 40 percent QE have been obtained. The dilute nitride work is now at the stage of material growth, with focus on how to reduce defect densities, improve crystal quality, and control background impurities. SEDD has achieved the highest incorporated nitrogen to date. On the IV-VI materials, SEDD is the only group in the world working on the growth of PbSnSeTe on Si. SEDD has reported the lowest defect density ever achieved in this material. SEDD has received funding from the Defense Advanced Research Projects Agency (DARPA) and other customers for the detector work. It has leveraged the expertise of various partners, including national laboratories (e.g., the Naval Research Laboratory and the National Aeronautics and Space Administration), universities (e.g., the Massachusetts Institute of Technology, Lehigh University, and the University of Illinois at Chicago), and industries (e.g., BAE Systems, Teledyne Technologies, and Raytheon). SEDD is involved in five Cooperative Research and Development Agreements, joint Army Technology Objectives, and CTAs.

Work on III-nitride materials is directed toward ultraviolet device applications in the wavelength range less than the cutoff wavelength of GaN. SEDD has state-of-the-art experimental facilities in this area, and very interesting data have been obtained. However, the potential for significant new insights into these materials appears to be hampered by limited theoretical support in the analysis of the results. Outside collaboration is not likely to be an adequate substitute for critical in-house discussion and analysis.

SEDD has been producing high-quality AlGaInN materials using molecular-beam epitaxy (MBE) and will be capable of growing this material using a new metal-organic chemical vapor deposition (MOCVD) system. Combining the MOCVD system with the MBE system, SEDD is able to produce various materials that can be utilized for optoelectronics and electronic devices such as high electron mobility transistors (HEMTs). From materials, to processing and fabrication, to packaging and characterization, SEDD is able to perform all of these functions in-house. This is important in being able to achieve its mission of fulfilling Army needs. SEDD has grown and fabricated AlGaN/GaN/InGaN light-emitting diodes (LEDs) operating at 340 nm and 280 nm wavelengths. The goal is to realize high power density emission from these LEDs. Some of the applications for these devices are water purification, biological agent detection, and non-line-of-sight communications. Avalanche photodiodes in both GaN and AlGaN materials have been fabricated for use in the visible-blind and solar-blind regions respectively. There are incipient activities toward the realization of lasers using nonpolar nitrides. For the LEDs, there were collaborations with the Palo Alto Research Center (PARC) on the MOCVD growth of III-nitride materials. This technology will now be transferred from PARC to ARL for use on the new MOCVD system.

There are many laboratories and centers worldwide investigating GaN devices and materials. SEDD is leading in the growth of nanoscale-compositional-inhomogenous (NCI) AlGaN materials with enhanced luminescence. The enhancement is due to localized high carrier density in the spatially nonuniform AlGaN. SEDD is also leading in the optical characterization of AlGaInN materials; this is a unique competency that is sought by outside collaborators. SEDD has received funding from DARPA, the Homeland Security Advanced Research Projects Agency (HSARPA), the Defense Threat Reduction Agency (DTRA), and other agencies. It has also leveraged the technical competencies of many collaborators, including Lehigh University; the University of California, Santa Barbara; and the Georgia Institute of Technology, along with industries such as Crystal IS and PARC. There is an ongoing collaboration with GE Global Research on ZnO. The investigations reported are among the best in the field and involve excellent staff. SEDD is working in a highly competitive area, and it has been able to maintain its prominence as a laboratory owing to excellence in personnel. As stated above, there are two clear areas—growth of NCI AlGaN and high-speed optical characterization—where SEDD leads the competition. The personnel in this group should collaborate closely with the RF group; the characterization technologies developed here will provide insights into HEMT materials and properties.

The Flexible Display Center (FDC) at Arizona State University (ASU) is a unique, first-rate program that complements what the others are doing in this area. The FDC encompasses research and development (R&D) and a pilot line for a key enabling technology for network-centric operations. The present focus is on silicon thin-film transistor technology for the display backplane, with the option of transitioning to organic field-effect transistors at a later date. The principal goal of the silicon work is to develop a process that employs only temperatures sufficiently low to be compatible with flexible substrates. The purpose of the FDC at ASU was made clear, and the advantages and issues were clearly articulated. The model for the industrial partnership, complemented by academic contributions, is an interesting new model of collaboration within the United States. The ARL-defined objectives were described and certainly will be extremely beneficial to the Army mission. It seems quite clear that the needs of the Army are defined and shared with the FDC personnel, and, it is hoped, with industrial partners (although this was not clearly explained by SEDD). A question that remains addresses the manner in which the ARL or Army needs are ultimately met, such as the need for robustness in extreme environments, lightweight displays, and appropriately low-power-consuming displays. Industrial partners include process equipment manufacturers in addition to the key materials and display corporations.

The Organic Light Emitting Diode group is attacking appropriate problems, generating intellectual property, and establishing a solid publication record. SEDD is strong in this area, and the work should be commended. The demonstration at the laboratory that works on organic materials for devices and displays highlighted the capability of the SEDD activity to include novel materials synthesis, materials deposition, device fabrication, and display manufacturing (on a research scale). The emphasis is currently focused on creating an efficient blue organic emitter, because this is technologically the limitation for full-color red-green-blue displays. A quantum chemist would perhaps be a valuable addition either to internal ARL staff or externally for collaboration on the project. The researchers are very enthusiastic with regard to discussing the research objectives and research accomplishments, and the laboratory demonstration was a success.

SEDD is working with the Institute for Collaborative Biotechnologies at the University of California, Santa Barbara, to develop microfluidics for the detection of anthrax, viruses, and other agents. This university collaboration appears to be working well, and the investigators appear particularly good at combing different techniques. The fundamental technology has very good potential for enhanced medical care. An E-DNA compact biosensor for chemical and biological detection was presented, with the goal of making a simple, multiple-use sensor platform for several different types of biological and/or environmental threats. The idea is to have a disposable chip that just plugs in to the personal digital assistant (PDA)-like platform. Deoxyribonucleic acid (DNA) is first separated electrophoretically on a small scale using microfluidics. The challenge is loss of material due to the small size of the DNA, which results in false negatives (noise); that is why there is a focus on doing the microfluidics well. Polymerase chain reaction is used as an effective amplifier of DNA, with the results being interrogated on a gold electrode. There are several electrodes on the chip, each with a different DNA type, and they are sequentially connected to the ground electrode. The SEDD work is different from other work in that this is an all-electronic sensor, made possible by microfluidics that are electronically controlled (no exterior valves, pumps, and so on). The electronics consumes around 1 W, mostly consumed by heaters. It is to the Board's knowledge a smaller platform than any other for this type of application.

Corrugated Quantum-Well Infrared Photodetectors

SEDD is the leader in corrugated quantum-well infrared photodetector technology. The concept is brilliantly simple. The key technologist in this area is known internationally for his success in designing and building multicolor infrared detectors. The results are world-class, with a 0.02 K resolution. This is a very remarkable effort which demonstrates that a great deal of persistence can turn a promising idea into a competitive technology. The leader of this effort has been involved with QWIPs from the very beginning, and the current work on C-QWIP focal plane arrays appears to have yielded an approach that is competitive with HgCdTe focal planes in performance and will have significant advantages in cost. The C-QWIP research being performed in SEDD represents R&D at its best.

The pertinence of the C-QWIP work to Army needs further enhances the importance of this research. In this hierarchy it is recognized that HgCdTe sensors offer the best performance in terms of sensitivity and quantum efficiency but require cryogenic cooling and are extremely difficult to fabricate. Bolometers are lower cost and lack the sensitivity of HgCdTe sensors but operate at room temperature (using thermoelectric coolers for stabilization). They are ubiquitous on the battlefield, being used as thermal weapon sights on guns and driver's vision enhancements on vehicles. The promise of QWIP sensors (which also require cryogenic cooling) is enhanced performance over bolometer sensors, with substantially lower cost than that for HgCdTe sensors. These QWIPS with enhanced performance would satisfy a number of Army missions, such as large-area persistent surveillance. QWIPs can achieve a much higher pixel count (needed for wide-area surveillance) than bolometers, at a lower cost than that for HgCdTe.

However, QWIP sensors, until SEDD's recent work, had a very low quantum efficiency and could never reach their potential as a practical device for an Army application. The QWIP quantum efficiency was approximately 3 percent as compared to approximately 85 percent for HgCdTe. The reason for this is the conventional technique of coupling IR radiation into the QWIP layered structure. A reflective grating is used that does not efficiently couple the IR radiation into the QWIP active layer. The grating also forces narrowband detection as compared to the wide-bandwidth detection of bolometers and HgCdTe, which further lowers the quantum efficiency. To raise the IR absorption and the quantum efficiency, the pixel size is increased, which limits the sensor resolution. SEDD invented a new concept for coupling the IR radiation into the QWIP active layer; it entails placing an inverted V-shaped reflector around each pixel so that the IR energy is coupled directly along the QWIP active layer. This optimizes the quantum efficiency. The assemblage of V-shaped structures gives rise to the name corrugated QWIP, or C-QWIP. The resulting improvements are extremely impressive, with an increase in QE from 5 percent to approximately 35 percent with broadband coupling from 6 to 12 microns. SEDD has fabricated QWIPs focal plane arrays with 2048 × 2048 pixels and demonstrated lower-cost cameras with a 1024 × 1024 resolution.

This is a stunning success and has brought QWIPs back into serious consideration for Army applications such as large-area persistent surveillance and helicopter piloting. The most dramatic improvement is the increase in quantum efficiency by over a factor of seven. The corrugated technique also permitted smaller pitch sizes and larger pixel-count arrays. Equally impressive with these improvements is the extensive theoretical modeling used by SEDD to support these advances. Often a clever idea like this is implemented without the supporting theoretical analysis. In these cases, while a major improvement can be demonstrated, it can never be fully exploited using a trial-by-trial approach. SEDD, however, has developed a series of integrated models to fully explain the C-QWIP performance. Some of the more important include a wavefunction model of the lattice to predict the molecular absorption spectrum, an electromagnetic field simulation of the IR coupling to the active layers, and a reflectivity optimization model including the effects of surface plasmons at the Au interface.

The combination of the breakthrough corrugated concept, extensive and comprehensive modeling, and outstanding experimental results defines this work as among the best in its field. Indeed no other group has equaled this performance. This work should receive the highest level of support from ARL. It is being deployed with other government agencies. It would behoove SEDD to consider the next focal plane development to be 1920 × 1080 resolution, to take full advantage of the commercial high-definition television equipment including digital video recorders and displays.

SEDD's C-QWIP work is so good that perhaps it should be considered a national asset. This is, of course, an endorsement of the ARL concept underlying its semiconductor fabrication facility: "Build it and they will come." Certainly, having such a fabrication facility at ARL acts as a magnet for researchers, and the Army benefits from the resultant outstanding work. It would be expensive to technically transfer the C-QWIP process and fabrication technique to any of the few remaining GaAs fabrication facilities left in the United States. If legal considerations permit, ARL management should consider an arrangement that would allow the fabrication of C-QWIP arrays at ARL to be purchased by camera manufacturers such as FLIR, DRS Technologies, BAE Systems, Insight, and others.

Power and Energy

SEDD is doing an excellent job of framing the research questions that need to be answered for high-energy batteries. The team is well respected and is doing high-quality work. The right scientific and technical issues with respect to a technical objective for the mission are being posed. Portable power is critical to soldiers, vehicles, and sensor applications, and SEDD's program is on target in terms of identifying key scientific and technical challenges for Army-specific battery needs. Although lithium (Li)-ion batteries are available commercially, the specific needs of the military are often different from those of the consumer market, and for this reason it is important to maintain a significant technical effort in this area. SEDD has a strong program in battery technology, reflected in the competence of the staff and the quality of their work. The Li-ion and other battery-related work done at ARL compares favorably to similar work being conducted externally. The battery field and that of Li-ion in particular are fairly crowded due to commercial success with batteries. However, this is a technology that is almost exclusively produced by Asian companies and thus warrants a U.S. research presence. Batteries used by the military often have environmental and safety considerations that are significantly different from those for consumer batteries. This is reflected by ARL's work in low- and high-temperature electrolytes, as well as by the safety testing of battery packs of different cell chemistry punctured by ammunition rounds. The SEDD effort is well respected in the field, and it is anticipated that the quality of work will continue in this vein. In the Li-ion field some degree of collaboration with a manufacturer is usually necessary in order to gauge the value of internally developed technology. This is so because the charge/discharge cycle life and/or safety of the system is invariably affected by any changes in the cell chemistry, and academic laboratories are not equipped to produce prototype batteries that can reliably test such parameters. The work of SEDD in electrolytes and cathode chemistry is of a high caliber and may lead to improvements in Li-ion chemistry aligned with military needs.

Isotope batteries for embedded sensors comprise an area of interesting work that is important to the Army. The concept here is to meet the need for batteries with extremely long life in isolated locations through the technology of isotope batteries. The underlying idea of collecting the charge emitted by radiation has been communicated through basic undergraduate instruction in modern physics and through the public media's treatment of NASA's use of the technology. The nuclear fission industry has done extensive and detailed research on this topic for more than six decades. NASA and the Idaho National Laboratory have the technology well calibrated—having, for example, complete charts of which isotope material to use for what duration of battery life. Batteries with a long shelf life and batteries with a long operating life are definitely a need for the Army. The needs are unique enough to justify doing research in a number of technologies. One type is isotope batteries; another is thermal batteries. There are commercial batteries currently in use in the public utility industry that approach a 20-year life. SEDD is doing important work and is enthusiastic about it. Beta-particle batteries are unlikely to be manufactured commercially because of the liability associated with radioactive sources. Applications definitely exist and will probably proliferate further. Maintaining some expertise in this area is necessary.

SEDD is alone in the world in building SiC thyristors with pulse power capability. SiC is a very promising technology in which SEDD is on the leading edge. SEDD has been conducting research on SiC applications and issues for quite awhile, and the investment has given it a significant advantage in bringing this technology to the soldier early. ARL employed the now commercially available SiC diode at a very early date. In the case of the metal-oxide-semiconductor field-effect transistor (MOSFET), there is great commercial interest and potential. Industry is very interested in the SiC MOSFET and will track its development closely. SiC thyristors are a niche product. They form the switching element of the electromagnetic (EM) armor and the rail gun. No other semiconductor component can handle the current pulses anticipated in such applications. Thyristors in general are considered a niche technology, being obsolete for most applications and existing primarily in legacy equipment and certain niche applications. A good example of a niche outside ARL is high-energy physics research. SiC has taken much longer to develop than expected for a semiconductor technology, much to the disappointment of its proponents. However, SiC has a great deal of promise for providing exceedingly lighter, faster, more powerful, and more capable power supplies and high power equipment. Thus SEDD's investment and leadership role will pay off handsomely as the technology continues to develop. Persistence and patience will be required to maintain SEDD's primacy, as breakthroughs and useful developments, though yielding high payoff, will continue to be difficult to achieve and will likely to continue to arrive all too slowly. Very good work in this area is published and presented by SEDD at conferences.

The specific application at hand is a leading-edge concept—using the SiC (and Si) gate turn-off thyristor (GTO) as a means of accelerating switching recovery time. Thyristors are notoriously slow to regain their ability to block voltage after commutation. By actively removing charge from the gate region, a useful capability of the GTO, recovery time is halved in the experiments that were shown. The demonstration was very well done. The conducting of the test seemed good, including discharge source, instrumentation, interruption of current, and high-voltage restore at the end to verify recovery voltage. This is reminiscent of synthetic testing used for high-voltage circuit breakers by industry leaders. This experimental validation is a significant development. It is applicable to semiconductor-based protection on high-voltage circuitry as well as to the EM armor and rail gun applications that were shown. Using the GTO, a technology that the Japanese have led in development for 25 years, represents an excellent capture of foreign technology for a U.S. application. Close collaboration with U.S. manufacturers made this happen. Such developments are encouraging, showing a significant advance that only one in a leading position could identify and exploit. Determining appropriate data on reliability is going to be necessary as SiC devices become more available and begin to appear in Army hardware. ARL lacks the resources to do this as it needs to be done—it is just too expensive. However, ARL's partnership with the major domestic manufacturers and its leading research position should encourage the manufacturers to be aggressive in getting data on reliability.

Nanocrystalline magnetic materials for direct current (DC)-DC power conversion is a key technology for hybrid electric vehicles and pulse power. These devices must be bidirectional, and for military vehicles they typically convert between 300 volt DC batteries and a 600 volt DC bus. These need to be efficient and, most importantly, need high power density (kilowatts per liter) (to a lesser extent high specific power [kilowatts per kilogram]). A second key design constraint is heat removal, which is particularly challenging in vehicle applications, which involve high power and severe volume constraints. The group at SEDD has established a working relationship with Carnegie Mellon University, the University of South Florida, and Magnetics, Inc., to develop high power density devices. A demonstration focused on inductor materials that are used in DC/DC converters. The hardware was an inductor mounted on a cold plate with liquid cooling. Thermal images (IR camera) provided a map of the core and winding during operation. These magnetic materials need to allow high magnetic saturation fluxes and low power losses at the desired frequency. Iron-based materials from Magnetics, Inc., were used. These materials are coated with polymers to reduce eddy currents and are particularly suited for high frequencies. The target application is the Future Combat Systems (FCS), and there is collaboration with the U.S. Army Tank and Automotive Research, Development, and Engineering Center (TARDEC) on hybrid electric vehicle design.

Goals have been established (6 to 8 kW/l), and progress toward these goals is evident. Unquestionably, this technology is important for the FCS. The research at SEDD is excellent—of high quality and relevant. At the same time, many others throughout the world are developing similar technology for automobiles and buses. It is likely that the heavy-duty vehicles of interest to the Department of Defense (DoD) require higher power density than is needed by commercial light vehicles, but close attention to developments in the commercial (particularly overseas) sector are warranted. SEDD used U.S. industry to overcome the technical problem at hand and used a U.S. university for the advantage of its advanced knowledge and experience—a superb model for progress. This is but one in a sequence of such advances that must be made to bring hybrid electric vehicle technology to Army vehicles. The demonstration of advanced nanocrystalline magnetics was performed well. Appropriate issues of heat generation and heat sinking and how they were incorporated into an effective and innovative design were illustrated quite capably.

The next step, already begun at SEDD, is to develop and apply high-energy capacitive storage. U.S. industry has an interest and some investment in ultracapacitor storage. The automobile companies will bring it into their products in the next few years, but the automobile companies will not take it far enough, because Army vehicles have a need for more power and energy than are required for commercial automobiles and light trucks. When the automobile companies have confidence in the technology, they will send it to universities, such as the University of Wisconsin and Virginia Polytechnic Institute and State University, for validation, as they have done with magnetics already. Advances will also appear in Europe and Japan; SEDD has shown the capability to capture such advances in electric power and energy from non-U.S. sources. SEDD is at the leading edge of nanocrystalline magnetics; it needs to catch up a bit in energy storage. There will be other technology issues in power and energy applied to Army vehicles. SEDD has a good model for success that will serve it well. These are high-payoff activities, and SEDD has shown the ability and engaged the people appropriate for doing a good job of following through with them.

Radio Frequency and Electronics

SEDD has a vision for advanced radio-frequency technologies that includes multifunction RF systems for future battlefield platforms to enhance lethality, survivability, and mobility. To implement this vision, the directorate is focusing on antennas and RF front ends, nanoelectronics and MEMS, RF sensors, prognostics and diagnostics, and RF-directed energy.

In order to improve antenna designs, in situ antenna modeling is being used to analyze antennas in the environment in which they will be used. SEDD is using rapid prototyping and fabrication, together with modeling and high-fidelity measurements, to design and demonstrate integrated antennas for Army applications. Examples shown were helmet-mounted antennas, in situ antennas for ground vehicles and unmanned aerial vehicles, lapel-mounted RF identification tags, and antennas designed to be worn by the soldier.

Antenna modeling for so many different applications is an extremely difficult problem, and it is useful to have multiple types of code centralized in one laboratory with people who are experts in knowing what type of numerical method to use for different situations. This effort includes numerical EM for antennas, which is a well-established and very heavily researched field. Well-known commercial codes are used, which is entirely appropriate. It was not clear how the choice of tool is made and why in some cases one tool might be better than another. Experienced researchers in this field of numerical EM for antennas tend to use their favorite code or method, mostly because they have it or they know how to use it. It would be a worthwhile effort to develop a methodology for what code to use for optimal antenna design in specific situations: real ground, finite ground, complex dielectric bodies, small bandwidth, broadband, single polarization, and others. An interesting matrix could be created that would be quite useful to antenna designers.

There are some relatively recent research papers that would be of interest to SEDD, although there are no commercial codes using the techniques described. For example, recently the IEEE Microwave Theory and Techniques Society's Microwave Prize was awarded to a numerical EM paper which showed that finite element modeling (FEM) can be done with large-domain elements very efficiently. Such a method might be well suited to the problems that the Army is faced with. SEDD is proposing that rather than develop new numerical methods in an already crowded field, it is better to develop the know-how related to the design of antennas in complicated environments.

Antenna modeling and simulation is a research project that can provide great benefit to the Army and can lead to improved front ends. If a more general analysis method were established by SEDD, the conclusions would be of interest to people in the field who do not wish to make numerical EM their expertise but need to know how to improve antenna designs. The SEDD group has excellent expertise in antenna design. Improving the approach over time to move from discrete antenna solutions to a more general systematic approach as a final goal would provide great benefit to the community at large.

Millimeter-Wave Imaging

SEDD is developing advanced technology for millimeter-wave (MMW) imaging. This approach offers solutions to imaging in adverse conditions where other imagers are impaired, such as when looking through dust for helicopter landing under brownout conditions, looking through fog and smoke, and for other applications such as concealed-weapons detection. SEDD has a clear objective of increasing resolution while decreasing size, weight, and cost. It also has a good transition plan through the Army's Communications-Electronics Research, Development, and Engineering Center (CERDEC) and its Aviation and Missile Research, Development, and Engineering Center (AMRDEC).

SEDD has made excellent progress, with significant accomplishments that extend the state of the art in MMW imaging. These accomplishments include extending depth of field using cubic phase elements; demonstrating broadband antireflection gratings and three-dimensional rotating beams for ranging; and measuring attenuation effects in high-density dust clouds. SEDD's approach consists of using a low-cost focal plane array with MMW lenses, which enables the demonstration of flat-panel MMW imaging. Combined with a novel antireflection grating and rigorous EM modeling and computation imaging, this approach may enable the next generation of MMW imagers that can be used in the field. SEDD is at or beyond the current state of the art for this technology. It is effectively leveraging the directorate's well-known expertise in microwave and millimeter-wave technology in a new area that holds great promise for meeting immediate Army needs.

An interesting extension of this project would be to evaluate the possibility of integrating MMW imaging with IR imaging to provide a display that incorporates both. SEDD has the expertise for both types of imaging systems. An integrated display using infrared detectors, such as the QWIP discussed earlier, and MMW detectors could provide an imaging system with performance that far exceeds that of any other system currently available.

Microelectromechanical Systems

The Microelectromechanical Systems Technology for MicroRobotics group has done an outstanding job, especially given the relatively low level of internal investment. The group has obtained DARPA financing for the piezoMEMS and nanomechanics work, and it has a well-planned top-level roadmap for FY 2005-FY 2010. It is competitive with its peers in the nanoscale technology area, and the group members are excited and energized regarding the potential for this new field. The overarching vision presented for microsystems was that of a scorpion-like, bio-inspired, biomimetic mobile sensor platform in the centimeter size range. Although this vision is extremely aggressive, the SEDD research teams have organized a comprehensive and complete set of research projects that have a strong likelihood of positive research results. Certainly, should such a device be made to work, it would be a disruptive sensing technology in much the same way that the mobile antitank mine developed under a DARPA program was disruptive to mine-clearing capabilities of the enemy. The ARL team has not yet thought through all the scenarios for the use of such disruptive mobile sensor platforms. However, it has taken the approach of making an aggressively early demonstration of subcomponents of the system, in particular a totally functioning, insect-scale, piezoelectric-actuator walking leg. Although the issue of supplying onboard power will be problematic, the SEDD team has done a thorough job of evaluating and minimizing the total power required. Encouraged by the DARPA Microsystems Technology Office, the SEDD team is developing the materials that may be used to sell a program of this sort to DARPA management and provide funding not only for ARL but for others in the field. These activities are strong indications that ARL is pursuing the right scientific and technical issues for microsystems applied to miniature mobile sensor platforms.

The ARL program for microsystems is conducted by a SEDD team that understands the underlying science and comprehends other, comparable work done in the field. There is not, to the Board's knowledge, any such program in any other part of the Army. The SEDD team is fully leveraging DARPA initiatives in the field and is exploiting all the expertise in lead zirconium titanate (PZT) piezoelectric actuator thin-film deposition in the Army. The team is aware of the microsystems (miniature robot) work at major universities such as the University of California, Berkeley; the Massachusetts Institute of Technology; Stanford University; and the University of Michigan. The SEDD team is leveraging its connections to others in the field for access to the necessary fabrication tools not available in ARL. Its acquiring of its own hydrofluoric vapor etch machine, which has increased yield and increased fabrication progress immensely in the past year, is commendable, as is the addition of the advanced HF etching system for MEMS fabrication. The shortened cycle times and improved yields will benefit all MEMS activities.

The work presented by the SEDD team in microsystems is at the state of the art—in particular, its efforts in micro- and nanoenergetics, microrobotics, and microswitching of RF. Its work on PZT, three-dimensional circuit elements, piezoelectric actuators, and microrobotic components is leading the field. The team's work on micro shock sensors, micro fatigue testing, and micro energy harvesting is near to, but slightly behind, the state of the art. Since much of the work that is at the state of the art is relatively new, it is understandable that the number of publications is still lower than desirable. However, there is a strong effort to develop patents (several in PZT and PZT actuators are in process), and the group of predominantly young engineers seems highly motivated to publish. Overall, the group seems strong, competitive, energetic, enthusiastic, and diligent, though recently formed and including young members.

It is performing well for its size and newness. It will be increasingly important for the microsystems team to continue maturing and to increase the rate of publication.

The program focuses on microsystems that are bio-inspired and is capitalizing on their in-house strength in actuation and MEMS activities. Attention to overall objectives is driving the application of system analysis to all endeavors, and issues related to ground mobility, motor and behavior control, and power subsystems are considered together to advance a demonstration in the near term. The near-term goal of demonstrating a robot is challenging, but it will certainly be instrumental in forcing all technological thrusts to contribute necessary project-directed components. The system design focus has advanced to consider power consumption as a function of time by looking at off-the-shelf technology in the near term, with attention to research-driven and advanced technology toward the longer-term goal of partial operation.

It seems clear that the demonstration and bio-inspired robot project will lead to fundamental advances in microelectromechanical and nanoelectromechanical systems (MEMS and NEMS). It is important that the overall system approach be applied to push forward the technology and research directions. The funding from DARPA is important to leverage the ARL monies dedicated to this task. Furthermore, the opportunity to use the leading-edge technology held at ARL in piezoMEMS is a great advantage for propelling the project forward rapidly. One point to note is the current movement to remove all lead from integrated circuits because of environmental concerns; the impact of such restrictions should be considered as the piezoMEMS technology is employed.

The MEMS work within SEDD on using such devices to improve the thermal coupling of power electronics is very good. It represents a novel technique for increasing thermal coupling beyond traditional indium soldering to heat sinks or microchannel coolers. For example, although indium-soldered microchannel coolers have been used for high-power laser diodes, the use of MEMS and MEMS bonding techniques offers several significant benefits in mounting laser diodes to a heat sink. The thermal resistance from the laser diode to the heat sink will be reduced and more repeatable, thereby allowing the operational temperature of the laser diodes to be reduced and more controllable. With lower and more controlled operational temperatures, the laser diodes are much less prone to premature failure (i.e., reliability problems). Alternatively, with the higher heat fluxes possible owing to a lower thermal resistance from the laser diode to the heat sink, the output power of the laser diode can be increased. The problem of reflowing of the indium solder that is used to attach the laser diodes to the heat sink can be eliminated, thereby allowing a preeminent failure mechanism of high-power lasers to be avoided. This work has been recognized by DARPA, and the MEMS effort should continue to receive ARL support.

Radar

The human radar signature investigations at ARL started in 2006. The goal is to find Doppler radar signatures related to human behaviors such as walking gait, breathing, speech, carrying heavy objects, changes when nervous, and so on. Since this is a new project, there is not yet too much progress, but it is an interesting research topic. Since radio frequencies are measured that reflect changes in the movement of the human body, including movement of the internal organs, physicians are also contributing information to further correlate to the obtained RF data. Significant progress was made in preparing the documents to obtain permission for research on humans. Such applications are extremely long and tedious to complete, but they are certainly important and valuable. The effort appears to have clearly considered the objectives and to have created the necessary collaborations to complement the internal ARL activities.

The large challenge will be in collecting a solid data set and then doing appropriate data processing. It seems that the group working on the project will need a very low phase noise local oscillator and long integration times for low Doppler shifts. For the heart muscle, velocity is 7 to 15 cm/s, which is a very low Doppler shift. The group is examining radar signatures from 200 MHz to approximately 100 GHz, but it was not clear yet what the best choice of frequency would be. Electromagnetic models and radar measurements were used to examine the polarimetric and Doppler signatures of a human body from ultrahigh frequency (UHF) through Ka-band frequency. The group is aware of work done at the University of Hawaii and by a company in the Netherlands. The human radar signature is a new project that started in 2006, with some promising initial work. ARL is aware of other work in this field, and the Board is not aware of any additional radar-related work that the SEDD group did not mention. This is a worthwhile research topic with some conclusions to be reached within a few years and seems like a constructive and appropriate research project with potentially interesting results.

Ultrawideband (UWB) penetrating radar for sensing through walls uses extremely short pulses to provide accurate range resolution. It is not a new idea, but it has only recently become practical with high-speed electronics. UWB radar has the potential to image through walls and other objects with low to modest electrical conductivity. Being able to do so would provide clear military and law enforcement advantages. The short pulses associated with the wide bandwidth allow for very accurate determination of distance. Multiple sensors or synthetic aperture techniques can be used to provide imaging.

ARL has experience with UWB radar. As early as 1995, ARL explored the military potential of UWB to penetrate foliage to find targets. The current work is focused on urban warfare and the technology necessary for finding personnel in buildings. This problem is different from that of the short-range UWB radar that is being pursued commercially and by the military for mine hunting. The urban warfare problem requires that meaningful images be generated by sensors operated at longer ranges than are required for most other applications and moving target indicator (MTI) capability. Under urban warfare conditions, the processing required to generate meaningful images is complicated not only by the changes in target and clutter reflectivity over the wide signal bandwidth, but also by the low signal-to-noise ratio of the typical returned signal. To address these and other issues, the current ARL program is focused on modeling and simulation and on data collection in realistic urban scenarios. The results of these efforts are being used to help develop algorithms for synthetic aperture radar image formation and MTI techniques. This is a meaningful foundation for the development of an important military and law enforcement capability. For its potential to be realized, it needs to be coupled with a strong signal-processing effort that can take advantage of that foundation. The low-frequency UWB radar work is being done in collaboration with the Army CERDEC's Intelligence Information Warfare Directorate and the Office of Naval Research.

Image Enhancement and Understanding

Two image-processing projects enhance the resolution of a long-wavelength infrared uncooled imager, which allows for better missile performance or lower-cost sensors. The first project took advantage of the fact that the imager moved as a unit while tracking a target that retained its shape between frames. This allowed the development of algorithms that filled in missing pixels from one frame with those available in others. A computationally inexpensive way to estimate the highest frequency that should be amplified before noise dominates was developed as part of the project. A critical feature of this effort was finding an efficient approach that could be realized in tactical hardware. The work was done with real sensor data and generated excellent results.

The second project was a super-resolution effort also done on real data. Super-resolution operates by amplifying the higher special frequencies in an image that are attenuated owing to system issues. One price of this process is the amplification of noise. The super-resolution and deblurring algorithms are novel but not unique. That said, every signal-processing algorithm needs to be customized to the sensor and the environment, and this project solved a real problem. What matters is the degree to which these algorithms are tuned to the application and how robustly they behave in environments that the Army cares about. For instance, does the super-resolution work if the background of the image has significant spatial frequency content? A commendable useful collaboration was established with personnel from the Night Vision Laboratory of the Human Research and Engineering Directorate (HRED), the Naval Research Laboratory, the Army AMRDEC, and the National Institute of Standards and Technology.

Both efforts show an understanding of the sensors and work with real data and effective algorithms that meet the operating constraints of a military system. This is an excellent formula for success. Three papers have been published in *Applied Optics* since 2006.

Image understanding, or machine recognition of complex images, is of critical importance to the military, because asymmetric warfare and long-range lethality require autonomous and semiautonomous processing of imagery. The challenges of processing such images involve finding a suitable representation space, the development and application of models, the rejection of clutter, the use of context, the application of constraints, training, finding robust solutions, and the choice and tailoring of a classifier. SEDD has begun working a new set of classifiers and applications. Recent publications include an enhanced matched filter technique in *IEEE Signal Processing Letters;* an eigenspace separation transform in *Geoscience and Remote Sensing Letters;* and a change detection method in the *Journal of Applied Remote Sensing*.

Sensing

The autonomous sensing activity at SEDD includes both sensing and data analysis. The work in acoustic sensing has a long history, and this group continues to play a leading role in the field. The results have led to fielded technology with a significant impact on Army operations. The activity in magnetic and electric field sensing is interesting, with clever engineering behind it. Regarding the "autonomous" part of autonomous sensing: the nature of what is meant by that term is changing rapidly in the sensing and signal-processing communities. While much is still speculative, there appears to be a convergence of technologies in sensing, signal processing, robotics, and networking, leading to an envisioned system of autonomous sensors making decisions about data collection in a feedback loop based on the analysis of previously gathered data. ARL's vision for autonomous sensing is not yet at this level, nor is it clear that it needs to be. The application of existing signal-processing and data fusion methodology to unattended ground sensing is clearly an important area to the Army, and that work is commendable. It is recommended that there be more interaction between the autonomous sensing group and the MEMS microsystems group. The small bio-inspired device may turn out to be precisely the kind of platform appropriate for tomorrow's highly mobile networked autonomous sensor.

The work in sensor fusion and image understanding is focused primarily on adapting extant technologies to Army needs, including immediate operational needs. SEDD does outstanding work in this area and has state-of-the-art capability. The aerostat approach is particularly effective. Significant work has been undertaken in applications of sensing technology to sound source detection. While it was difficult to distinguish between the research contributions of the SEDD researchers and those of the group's contractors, the group discussed its research in depth and professionally, and the quality of the research is quite good. One important piece of work is the acoustic propagation modeling. Other acoustics topics presented were infrasound and vehicle tracking. Collaboration with other in-house organizations (e.g., CISD) and outside organizations could strengthen this activity.

The acoustics group continues to be active in the Military Sensing Symposia on Battlefield Acoustic and Magnetic Sensing (MSS BAMS), SPIE meetings, and publishing in the proceedings. In November 2008, the group helped sponsor a special session at the Acoustical Society of America meeting in Miami, Florida, on acoustics for battlefield operations and homeland security. The group continues to play active roles in the Long Range Sound Symposia. In June 2008, the acoustics group participated in the International Technology Alliance NATO measurement exercise in Bourges, France, to localize impulsive battlefield sounds from military sound sources using multiple sensing platforms.

The unattended ground sensors group effort is acquiring an active frequency modulated sonar unit from the Johns Hopkins University Applied Physics Laboratory that will allow ultrasonic Doppler measurements in conjunction with other sensing technologies including IR and visible technologies for the human factors research. This work should complement the similar micro-Doppler radar work for human motion that was initiated in 2008. The group should review the open literature on human-cadence-detection signal processing and consider the possible incorporation of other sensing technologies within the group, including the electric and magnetic field sensors and IR technologies to sense human motion behavior.

The acoustics group continues to devote significant time to technologies deployable in the field in the near term. Its contributions to measurement and signature intelligence (MASINT) applications are notable, including contributions to improving the microphone performance of the unattended transient acoustic MASINT sensor (UTAMS).[2] From an acoustical perspective, the investigative approach is analytical and interesting.

Low-cost unattended sensors to monitor power-line usage, vehicle movements, and other activity demonstrated, and information was presented on electric field sensing, including information on vehicle signatures in power-line electric fields; passive, remote classification of power-line activity; and underground electric field sensing (resistivity imaging). The sensors in all of these applications can be very low cost and can have long lifetimes, making them suitable for extended surveillance. The realizations being pursued are applicable to important, current military needs. This work requires very sensitive measurements of low signals in the presence of much larger signals, and it needs to be done with low-cost sensors. In some cases it requires fusion with magnetic, acoustic, or seismic sensors. The work is outstanding, including field data collection, modeling, analysis, and a unique laboratory measurement capability. Collaboration with a commercial company allowed for the production of low-cost sensors, and other collaborations provided specialty skills.

OPPORTUNITIES AND CHALLENGES

Overall, the Sensors and Electron Devices Directorate is performing at an outstanding level in the research, development, and deployment of technologies that have both near-term and long-term benefit to the Army.

The breadth of the SEDD work in IR detectors is very impressive. In some cases a critical assessment is perhaps necessary to determine if sufficient resources are available to pursue all of these efforts

[2] MASINT is scientific and technical intelligence information obtained by quantitative and qualitative analysis of data (metric, angle, spatial, wavelength, time dependence, modulation, plasma, and hydromagnetic) derived from specific technical sensors for the purpose of identifying any distinctive features associated with the source, emitter, or sender and to facilitate subsequent identification and/or measurement of the same. UTAMS is an acoustic sensor system created by ARL, used to locate sources of hostile artillery and improvised explosive devices.

simultaneously. Specifically, this may be an issue with the Type II superlattice work, where state-of-the-art results in the mid-wavelength infrared (MWIR) regime were achieved, but the extension into the LWIR regime appears to have made little progress. The continued examination of the Type II superlattice approach is possibly less productive and resulting in a dilution of other effort. SEDD has been responsive to feedback from the Board in this area.

A well-known contribution of the SEDD acoustics group is its continued and significant presence at the MSS BAMS. The SEDD acoustics group plays the lead role in organizing this meeting and presents research papers on current SEDD acoustic research efforts. The group is also active in other meetings, including the NATO SET 107 and the SPIE annual conference in Orlando, Florida. This being said, two weaknesses of the acoustics research group are the lack of acoustic publications in refereed journals and poor or irregular attendance at professional society meetings such as those of the Acoustical Society of America (ASA). Clearly, during wartime ARL's focus shifts from the more basic 6.1 and 6.2 research to engineering development. This was quite evident in several of the posters presented, and also evidenced at recent MSS BAMS meetings. However, for ARL to acquire and maintain its standing as a prestigious acoustics research group, acoustic research that is publishable in peer-reviewed scientific journals must be accomplished. This is not occurring now. The ARL scientists conducting acoustics research are outstanding contributors and should be encouraged to attend the professional society meetings that are closely tied to the peer-reviewed journals on a regular basis. An example is the *Journal of the Acoustical Society of America* and the ASA meetings, where there are routinely sessions on outdoor sound propagation and the coupling of airborne sounds into the ground. Such activities require recognition by ARL management that time is needed by journal authors to write and even rewrite manuscripts after the review process. Also, since HRED has the new Environment for Auditory Research facility, collaboration should be encouraged between the SEDD acoustics group and HRED.

During the next assessment, it would be helpful for SEDD to further elucidate how ARL supports the other branches of the Army in terms of portable power needs, how the various branches collaborate, or how advances in battery chemistry are transitioned from ARL to deployed products. This would enhance the Board's understanding of the extent to which the research program reflects a broad understanding of the underlying science and of comparable work being done within other ARL units and within the DoD, as well as in industry, academia, and other federal laboratories, and how well it employs the necessary resources with respect to instrumentation and other elements. The demand for light, compact power sources in the U.S. military is large and growing. The importance of this technology warrants a well-coordinated response in the military laboratories, and this appears to be lacking. There also needs to be some further discussion of how ARL intends to move technology from the laboratory to the soldier. This is particularly true in the United States, where there is no significant manufacturing base for Li-ion chemistry.

ARL management should consider strategies for formalizing specific management and reward structures for crosscutting projects that encompass teams of researchers drawn from multiple directorates. There is a particular opportunity for these types of collaborations at this time, since CISD is beginning an effort in multicore processors and embedded supercomputing. These projects need effective applications, and there are several SEDD projects that clearly need the computing horsepower. Together, these need to be recognized as systems projects with cross-directorate ownership. SEDD could also benefit from an expanded focus on data-analysis techniques, specifically machine learning and data-mining algorithms, and from incorporating these methods into the system designs. This can be accomplished with additional staff focused on system design and data analysis or by building collaborations with personnel in the CISD. Another example would be collaboration with personnel in CISD known to have strength in signal processing.

OVERALL TECHNICAL QUALITY OF THE WORK

SEDD management is doing an outstanding job in the following areas. It has been successful in attracting new talent to the organization and retaining the best of the existing staff. SEDD management has created a dynamic environment for creative research, and staff morale seems very high. SEDD has developed a strategy to build a top-notch and in some cases unique infrastructure as a mechanism to attract outside collaborators. This is an excellent approach that seems to be working. In the long term it will be of significant benefit both to the Army and to the scientific community as a whole.

A strong and commonly held culture was observed with regard to SEDD's "owning" science and technology on behalf of the warfighter. From a science perspective, SEDD owns the applications and actively seeks to work with the best scientists to innovate new or better solutions. The staff shows a generally impressive understanding of both the applications and the relevant science and in-depth understanding and eagerness to provide the warfighter with an improved product. Dedicated staff is working in a very good infrastructure. Attracting and retaining the staff with the Army's needs as their motivation is the key deliverable for those who are stewards of ARL. On that front, SEDD is in great shape. The leadership should consider ways to increase the odds that some of the new talent added over the past 5 years will stay at ARL and will have an impact. In addition to the physical infrastructure, the "brain trust" needs to be kept current. There is a great mix of junior and senior staff members within SEDD.

There is also access to a wide range of industry and academia. The interaction between SEDD personnel and outside collaborators in industry and academia is commendably strong, and interactions between ARO and SEDD staff should be expanded. Based on the work presented, SEDD is extremely strong in the projects that may be characterized as engineering or systems development. SEDD has demonstrated that it can make contributions that have immediate impact on current or near-future Army operations. The scientists and engineers involved in this work are highly motivated and goal-oriented. Novel device and system work, particularly the kind that requires the integration of a team of researchers spanning multiple disciplines, also is executed with much energy and enthusiasm.

There appears to be a trend in SEDD to more applied research as opposed to basic research. Given the immediate demands on the Army, this is understandable, but it is hoped that the trend will be self-correcting over time so that ARL does not lose its focus on basic, 6.1-type research. SEDD does appear to have a healthy ability to change research directions over time based on Army needs, with 10 to 20 percent of its projects being redirected yearly.

In the future, the Board would like to see the work that aims at fundamental materials research that supports the more applied work. A strong theoretical support base can have a significant impact on the materials research within SEDD. Theoretical work provides guidance to some of the efforts and is indispensable for thorough analysis of the data generated.

The scientific quality of the research in SEDD is clearly of comparable technical quality to that executed in leading federal, university, and industrial laboratories, both nationally and internationally. This is not unique to a single area of SEDD but can be demonstrated throughout the directorate. For example, SEDD is a leader in infrared detector technology, advanced RF technologies, image processing, and power components such as SiC thyristors. The SEDD research program reflects a broad understanding of the underlying science, which is reflected in the overall success of the research and development programs. In general, SEDD strives to understand what other research teams are doing here in the United States as well as internationally.

The qualifications of the SEDD research team are compatible with the challenges of SEDD research. SEDD is fortunate to have a leadership team that has created an environment that attracts and retains top scientific and engineering talent. The technical community recognizes the excellence of these qualifications through awards and honors, such as IEEE Fellow status. Research challenges are best addressed not simply through staff credentials but through the combination of talent, motivation, and teamwork. SEDD scientists and engineers are not only talented—they are also highly motivated and work together in a team environment to deliver solutions to Army problems.

An organization such as ARL is judged primarily on the transition of solutions from the laboratory to the field. Thus, experimental results are usually at the forefront compared to theoretical results. SEDD does an excellent job of using computational methods to support experimental procedures, such as in signal processing or antenna design. Theoretical foundations are not necessarily given as much emphasis as computation or experimental methods. While SEDD has strength in the theory behind its research, there are some areas, as noted in the "Opportunities and Challenges" section above, where additional support in the development of deep theoretical foundations could benefit ongoing and future programs.

SEDD has well-equipped laboratories and facilities to support its ongoing projects and is responsive to the facilities needs of its researchers. As an example, the recent acquisition of new etching tools for MEMS fabrication has enhanced the production of these devices. That being said, acquiring the capital investment necessary for state-of-the-art research is an ongoing battle, especially given current and future funding constraints. The facilities and equipment available to SEDD determine in part whether the appropriate human resources will be available to achieve future success. Top-notch scientists and engineers are attracted to organizations that provide not only interesting projects but also the capital investment necessary to achieve the goals of those projects. To this end SEDD will require the support of ARL management to ensure that it has the investment resources necessary to keep SEDD at the forefront of research activities worldwide.

SEDD management and SEDD researchers have been extremely responsive to the Board's recommendations. The technical assessment process takes time and resources in order to prepare, present, and discuss reviewed programs and projects. SEDD has taken a proactive and positive approach to the assessment process, which in the Board's perception has become beneficial to all involved.

5

Survivability and Lethality Analysis Directorate

INTRODUCTION

The Survivability and Lethality Analysis Directorate (SLAD) was reviewed by the Panel on Survivability and Lethality Analysis of the Army Research Laboratory Technical Assessment Board (ARLTAB) during July 10-12, 2007, at White Sands Missile Range, New Mexico, and New Mexico State University (Physical Sciences Laboratory), and during July 22-24, 2008, at the Army Research Laboratory (ARL), Aberdeen Proving Ground, Maryland.

SLAD is the U.S. Army's primary source of survivability, lethality, and vulnerability (SLV) analysis and evaluation support with regard to major Army systems. SLAD's general objective is to ensure that soldiers and systems can survive and function on the battlefield and to assess the degree to which Army systems are reliably lethal to enemy forces. Its mission includes SLV analysis and assessment through the entire life cycle of major Army systems, from development through acquisition to deployment and operation, in the context of a full spectrum of battlespace environments and threat forces, tactics, and systems. SLAD further provides advice to Army Headquarters, program executive officers, and subordinate program managers, as well as an array of other evaluators, system developers, and Army contractors, and other defense-oriented laboratories. Finally, SLAD is tasked with supporting special studies and inquiries motivated by and affecting current military operations.

The SLAD portfolio comprises a very large number of relatively small tasks. It has been difficult for the Board to understand, until this cycle, exactly what proportion of the portfolio it is exposed to. Based on a work breakdown structure provided at the Board's request, it appears that the Board is able to examine something less than 20 percent of the portfolio per year at a level of technical detail sufficient to assess the technical quality of the work. The Board's emphasis has been on those tasks that show the strongest continuity of effort, that have the broadest influence on SLAD's overall performance, and that most closely fit the charter of the Board.

In contrast to most other directorates at ARL, SLAD's portfolio includes relatively little applied research funding and no basic research funding. The overwhelming majority of SLAD funding is later in the Department of Defense (DoD) research, development, test and evaluation (RDT&E) chain, either provided by acquisition programs for SLV support or by RDT&E management support funding organic to ARL. The small fraction of applied research funding supporting SLAD is devoted to the development of tools, techniques, and methodologies required to undertake SLV analysis and assessment. This portfolio of funding reflects a relatively long period of stable SLV techniques, emphasizing ballistic survivability of armored systems and lethality of U.S. weapons systems against armored systems. SLAD is now necessarily supporting SLV analysis in a much broader and more rapidly evolving context, in which communications, networking, and information processing, rather than weapons systems per se, are believed to be the essential and sustaining advantage of U.S. military forces. A central concern of the Board remains the issue of whether the directorate's funding portfolio will result in tools, techniques, and methodologies capable of providing the Army with the assessment capability needed under the emerging paradigm of network-centric warfare in an irregular battlespace. Tables A.1 and A.2 in Appendix A respectively show the funding profile and the staffing profile for SLAD.

CHANGES SINCE THE PREVIOUS REVIEW

The proportion of SLAD's efforts comprising special studies and inquiries motivated by current operations has increased substantially since the advent of Operation Enduring Freedom and Operation Iraqi Freedom. The Board has been exposed to many of these efforts over the recent assessment cycle, necessarily at the expense of the rest of the portfolio. Most of these efforts are of relatively short duration, and although the theme of special operationally oriented studies is clear, continuity of effort and methodological progress are not something that is easily visible to the Board. SLAD's contributions to the war effort have been competently performed, often under very serious time and resource constraints, and have apparently been significant influences on current operations and supporting acquisition. The Board has been exposed to a large number of these efforts and has consistently been impressed with the dedication and ingenuity of the staff involved. However, this work is *not* research per se and hence is difficult to evaluate in the context of industrial laboratory or academic research. In particular, it is extremely difficult to understand whether highly responsive and time-constrained analysis and engineering work is among the best in its field, since the field of comparison is necessarily limited.

As noted previously, the SLAD portfolio is very granular, and the Board can sample a relatively small fraction of the individual tasks supported by SLAD. SLAD management has tended to emphasize the operationally oriented tasks and special studies in developing agendas for the assessment meetings. For the future, SLAD should consider leaving the assessment of this work primarily to SLAD's operational customers, who are obviously in the best position to assess its impact and relevance, and to refocus the Board's attention primarily on methodological progress, tool and infrastructure development, and the development of necessary new capabilities. To the extent that the Board continues to be exposed to the operationally oriented studies, it will focus primarily on the degree to which SLAD is engaging the external technical community to leverage previous work (both academic and industrial) to enhance the technical credibility and quality of its products.

Beyond the change described above, SLAD has experienced other significant changes over the 2-year assessment period. One of the most significant of these was the establishment of Warfighter Survivability Branch within the Ballistics and Nuclear, Biological, Chemical Division. This branch was established to provide an organizational focus within SLAD for the soldier-focused portion of the survivability mission. The creation of the new branch brings together many of the disparate tasks supporting current operations; more importantly, it provides the potential for a future focus on soldier survivability in the context of longer-term acquisition programs. ARL and SLAD management expect SLAD funding to decrease in coming fiscal years. SLAD has continued to work on methodologies aimed at assessing the effectiveness of system of systems (SoS), which has been a continuing recommendation of the Board.

However, the methodological development has focused increasingly on the System of Systems Survivability Simulation (S4), a fine-grained, event-driven simulation whose development is focused on human decision-making processes. Methodological development focusing on the Mission and Means Framework (MMF) has essentially stopped; the MMF is an approach to decomposing missions and systems for analytically identifying links between subsystems and mission performance. Finally, in 2006, ARL management indicated that it was willing to consider a Strategic Technology Initiative (STI) in the area of SoS methodology under SLAD leadership. The Board strongly recommended that SLAD avail itself of this opportunity and, in particular, that SLAD add a third leg to its platform of SoS methodologies. This third methodology should provide enough fidelity to enable a meaningful study of scenarios for identifying major system-level impacts of, for example, communications bandwidth; intelligence, surveillance, and reconnaissance (ISR); and precision weaponry, without modeling fine-grain entities such as packet-level communications or details of terrain. Developing this methodology in collaboration with an extramural team other than the team that has been developing the S4 tool would stimulate needed fresh perspectives in SoS analysis.

In early 2008, a group consisting of members from the Panel on Survivability and Lethality Analysis and the Soldier Systems Panel met with ARL management on an STI proposal jointly prepared by SLAD and the Human Research and Engineering Directorate (HRED), and relying on S4 as the centerpiece of the approach. Qualified support was provided by the panel members for this proposal, which was not crisply defined. Progress on the STI was not assessed during this evaluation cycle, but the Board notes its disappointment that SLAD management declined to follow the Board's recommendation to develop an additional approach to SoS analysis.

ACCOMPLISHMENTS AND ADVANCEMENTS

Response on Improvised Explosive Devices

One continuing theme since early in Operation Iraqi Freedom has been SLAD's support of counter-IED (improvised explosive device) operations and acquisition. Beginning with the IED counter electronic (ICE) device system, SLAD has collected an extensive data set and developed significant capabilities that are operationally relevant. The demonstration of the follow-on system, DICE, in 2007, and especially the influence that SLAD had on the rapid acquisition of mine-resistant ambush-protected (MRAP) vehicles in 2007 and 2008, are impressive. The technical quality of the work in this area is good, especially considering the time and resource constraints under which it takes place. As noted in the previous section, however, the evaluation of this work is most appropriately done in the context of operations, not research and development per se. In the MRAP work, many engineering assumptions were made to simplify analysis (consistent with the time constraints), and while some validation was performed using live-fire data, there was no presentation of statistical analysis of data scatter or error budget analysis to address remaining sources of uncertainty.

Communications System Support

SLAD has a long history of providing excellent vulnerability support for the most widely deployed radios used by ground forces—the single channel ground and airborne radio system, or SINCGARS.[1] The breadth of the effort—starting from a focus on a very important issue and continuing with the depth of the analysis, the results, product improvements, the support for implementation, and rapid deployment—has been exemplary.

SLAD and the SINCGARS program manager have been working together to determine the performance of improved SINCGARS against electronic warfare (EW) threats and to ensure that the improvements enhance antijam performance through comparisons with previous versions of SINCGARS. SLAD should be commended for the development of state-of-the-art laboratory tools emulating current and emerging EW threat systems (CSAL), and state-of-the-art automatic generation of radio performance curves (CEWIS). Laboratory investigations of radios without technical publications require fairly extensive investigation with levels of reverse engineering. Today's technology enables fast and sophisticated EW threat systems; components are available off the shelf at very affordable cost.

SLAD work on SINCGARS provides an excellent basis for the directorate's becoming the leader in vulnerability assessment of the next generation of radios, that is, the Joint Tactical Radio System (JTRS). SLAD now has an encouraging agreement with the JTRS Joint Program Executive Office that it will get actual JTRS code (the Board expects that it will be for both soldier radio waveform [SRW] and wideband networking waveform [WNW]) and that it will not be continuing with surrogates or pre–Engineering Development Model JTRS code (e.g., SLICE). SLAD has also tried to use 802.11g as a surrogate; however, the orthogonal frequency division multiplexing (OFDM) in 802.11g differs from the OFDM approach in JTRS (e.g., for the WNW). This is especially important since results to date have shown that the issues previously identified and solved for SINCGARS have not been addressed in the pre-JTRS units that have been tested. If there are any further delays in getting actual code, SLAD can leverage other relationships to press for quick delivery and testing of actual JTRS code. Furthermore, considering the complexity and much broader application of JTRS as an Internet Protocol networked radio, the Board expects that there are many more vulnerability issues that need to be addressed than in SINCGARS or even SLICE.

Information Operations

SLAD achieved something of a breakthrough in communicating to the Board its support for Army information operations and assurance efforts during this assessment cycle.

SLAD participates in ongoing experimentation and demonstration with respect to command, control, communications, computers, intelligence, surveillance, and reconnaissance (C4ISR) on-the-move capabilities. It supported information assurance (IA) testing and determination of compliance with Army regulations and policies, and it was tasked with the analysis and identification of critical system and/or network vulnerabilities that may be exploited by an adversary, as well as the development of mitigation strategies for all system and/or network vulnerabilities. SLAD employed hacker methodology and conducted penetration testing. It has developed an impressive tool set in INVA/DE.[2] However, the

---

[1] SINCGARS is a combat network radio currently used by U.S. and allied military forces. The radios, which handle voice and data communications, are designed to be reliable, secure, and easily maintained. Vehicle-mount, backpack, airborne, and handheld form factors are available.

[2] INVA/DE is a tool set for performing audit and compliance testing against computer network attacks and computer network exploitation. INVA/DE consists of an integrated collection of public domain and SLAD-developed exploits/utilities resident on a

specific results or lessons learned and the extent of the vulnerability assessment effort for this large-scale experiment were not clear, since they were not explained. This effort presents another potential opening to establish a position to drive issues rather than to react to them. The experience gained should be leveraged to develop an overall network vulnerability assessment methodology and to define specific metrics to evaluate performance in this area.

Another area presented was SLAD's black core analysis. The black core concept is fundamental to building the Core Backbone Department of Defense (DoD) network. It represents the design of multiple logical networks and associated interfaces and includes encryption and agreements on specific information needed for quality of service and routing across the boundary to enable the transport of these logical networks over a single physical network. In January 2006, the Army Chief Information Officer (CIO/G-6) selected ARL-SLAD as the most qualified Army IA organization and designated ARL-SLAD as the technical lead organization for system-of-systems network vulnerability assessment of the next generation of networks for ground forces.

FY 2007 activities included modeling and simulation activities and vulnerability assessment of the Army's Future Combat Systems (FCS) middleware (SOSCOE)[3] source code analysis as well as an assessment of the FCS-proposed tactical public key infrastructure (PKI) and firewall analysis and risk assessment. However, the work that was specifically discussed was really an interface analysis to confirm or dispute other results (including FCS program manager and National Security Agency results), just barely impinging on the black core. This effort was not commensurate with the role assigned by the CIO/G-6. The methodology appeared to be quite simplistic, possibly because the black core design is not complete and the level of detail used in the modeling was necessarily coarse. While it is certainly useful to do a security analysis of an incomplete design, questions arise. In the opinion of the analyst, how much confidence should one place in the current analysis? Can the analysis be used incrementally as the basis for a more complete analysis when the full protocol is available? Although SLAD describes the risk of the design as low, what implementation details might cause the risk to increase?

SLAD should use this effort and participation in the 3Star Network Vulnerability Analysis meetings as an opening to address the overall security architecture, focusing on obtaining funding for the needed level of resources. SLAD should develop work plans that are broader in scope and commensurate with the role that the CIO/G-6 assigned to SLAD.

A highlight of the panel's experience during this evaluation cycle was the live demonstration of the information operations exploitation laboratory at White Sands Missile Range, which drove home the quality of the overall effort. A few recommendations might be useful. The main one (repeated from a previous evaluation cycle) is to consider the testbed interface software developed at the University of Utah's Emulab. That software is being used for a large, secure testbed at the University of Southern California's Information Sciences Institute under the name Deter Lab. The interface and associated tools are powerful and easy to use, and Deter Lab would be interested in working with ARL on spreading the technology. This might even obviate physically separating the exploitation laboratory from the testing laboratory. Another suggestion is to do more outreach on preparing materials that might be useful to red (adversarial) teams under contract to DoD. The Defense Advanced Research Projects Agency (DARPA)

---

laptop outfitted with ethernet and wireless network interfaces (Bluetooth, Wi-Fi, ZigBee, WiMax) with distributed capabilities.

[3] SOSCOE is the foundation for FCS networked software including vehicle management systems, C4ISR, and soldier and unmanned air and ground systems. Just as an operating system on a computer allowing one to interact with resources and other computers, SOSCOE allows battlefield systems to communicate and interact with the unit of action.

spends quite heavily on red teams, and establishing quality standards for red teams might be an area where ARL could help make a difference.

Another recommendation that the Board has made in the past is that SLAD keep informed by attending research conferences such as the following: the IEEE Security and Privacy Conference, the Advanced Computing Systems Association (USENIX) Security Conference, the Internet Society's Network and Distributed System Security Symposium, the ACM[4] Computer and Communications Security Conference, the Applied Computer Security Applications Conference (ACSAC), and others. Attendance at such conferences would keep SLAD aware of the trends in defense and detection. SLAD should also present a research paper at one of the conferences, and the Board recommends ACSAC as a likely venue.

System of Systems Survivability Simulation (S4)

The System of Systems Survivability Simulation (S4) has made very significant progress since the previous evaluation cycle. The development of additional tools should significantly enhance the productivity of analysis using S4; the increased focus on more concrete productivity enhancements and the de-emphasis of highly abstract and difficult-to-communicate approaches such as formal method analysis are commendable. The appropriation of the Menard graphical summary of force structure is a key example; this approach will be much more appealing and intuitive to the Army operational audience who will need to understand the S4 results. Similarly, the scenario-building capability will greatly increase flexibility in analyzing diverse scenarios (as will be required in an increasingly irregular context).

Several further developments will be required in order for S4 to fulfill its potential in supporting real decisions. These include validation of results, characterization of the fidelity of the platform and network information embodied in S4, and, perhaps most important, the development of design reference missions that are supported by Army leadership (probably embodied in the Training and Doctrine Command [TRADOC]). Validation and verification of S4 results are still at the stage of an operational expert rationalizing the observed simulation behavior. At a minimum, it is appropriate to expand the community of operators involved to include a more diverse experience base, particularly in the context of irregular operations. This may also be an appropriate point at which to formally engage TRADOC, which would facilitate closure on digital rights management (DRM).

The Board recommended in 2007 that it was also time to consider the implications of much more powerful computers than have been used to run S4. Although SLAD noted to the Board that S4 had been implemented on supercomputer hardware as of July 2008, it seemed that this development may have been a box-checking exercise. The Board emphasizes that the use of high-performance computing may afford significant opportunities to exploit the S4 tool. In other fields, established communities of computational experts have found high-performance computing to be a disruptive development, opening new frontiers that had not been appreciated because existing computational platforms were adequate to perform conventional analyses.

Finally, the Board notes that it previously urged SLAD not to commit exclusively to S4 as its principal SoS simulation tool, particularly if given the opportunity to apply additional resources from an ARL STI. First, although S4 has clearly advanced in utility and flexibility since previous exposures, there are still questions as to the self-consistency of its architecture, its robustness with respect to widely varying temporal and spatial granularity, and the extent to which sufficient runs can be both executed and analyzed to address broad SoS issues (particularly without the use of high-performance comput-

[4]ACM (Association for Computing Machinery) is an educational and scientific society uniting the world's computing educators, researchers, and professionals to inspire dialogue, share resources, and address the field's challenges.

ing). Second, SLAD's collaboration footprint has been a primary concern of this Board for many years. Developing another significant collaboration, along different lines of analysis and with collaborators other than those at New Mexico State University, could essentially double that footprint. That opportunity seems to have been lost. This is particularly troublesome now that organic SLAD resources are being used to support the S4 effort, instead of its being funded through the congressionally directed program under which it began.

Warfighter Survivability Branch

As noted above, the new Warfighter Survivability Branch provides an organizational focus for soldier survivability. The branch is actively engaged in the counter-IED efforts discussed above. It is primarily responsible for the Operational Requirements-based Casualty Assessment (ORCA) program, designed to develop a tool set to characterize soldier injuries and estimate casualties and performance degradation produced by enemy munitions. Finally, the branch is integrally involved in the Joint Trauma Analysis and Prevention of Injury in Combat (JTAPIC) program.

The JTAPIC program is led by the Army Medical Research and Materiel Command, with participation by the Armed Forces Medical Examiner, the Naval Health Research Center, the Institute of Surgical Research, the Aeromedical Research Laboratory, the National Ground Intelligence Center (NGIC), ARL, and the Program Manager for Soldier Equipment. The program combines medical, materiel, and operational intelligence data to improve the understanding of events that have caused casualties and to develop solutions that will mitigate future blast-related injuries.

The SLAD role in the program is to re-create the reported casualty-generating event by modeling vehicle configuration and crew positions, to analyze and model threat characteristics through reverse engineering, to compare predicted injuries and platform vulnerabilities with the actual data, and to examine potential mitigation techniques. In spite of uncertainties in the reported events and difficulties caused by different terminologies used in the diverse communities, it has proven possible in many cases to apply such SLAD tools as Modular UNIX-based Vulnerability Estimation Suite (MUVES)-S2 and ORCA to model actual events with sufficient accuracy to explain the observed injuries and damage. The expertise of ARL in processing and analyzing fragments recovered in the field is an important contributor to the program.

The JTAPIC program has accomplished impressive results under difficult time constraints. One case was described in which the feedback of analytical results was rapid enough to affect an ongoing operation, locating and neutralizing a specific threat. This performance was made possible by the close relationship of ARL to NGIC, and of NGIC to the operational forces. In another case, where damage to a vehicle led to casualties, the application of MUVES-S2 (including BRL-CAD[5] and ORCA) permitted ARL to show that the use of curtains would reduce the spread of behind-armor debris and improve survivability in similar future incidents. Analysis of locally devised armor, applied to some vehicles in the field, showed an actual decrease in survivability, leading to recommendations to avoid this practice.

The JTAPIC program is a clear demonstration that ARL, with its survivability tools and techniques, can affect not only future procurements of Army materiel but can provide rapid and valuable feedback to forces in the field. Compared with the situation encountered in the first Gulf War there has been a vast improvement in data gathering and subsequent processing, and ARL has been a key contributor.

[5]BRL-CAD is a constructive solid geometry solid modeling computer-aided design (CAD) system. (The acronym "BRL," for "Ballistic Research Laboratory," refers to the former name of SLAD.) It includes an interactive geometry editor, ray-tracing support for graphics rendering and geometric analysis, computer network distributed framebuffer support, image-processing, and signal-processing tools.

The analysis of data on all too many field incidents provides a unique opportunity for ARL to validate and refine its models.

The success of this program, and SLAD's contributions to it, offer a prime example of the benefits obtained by teaming between organizations with complementary areas of expertise. This is a case in which SLAD overcame its insular tendencies with significant results for its customers and similarly significant professional development of its staff. Extending the domain of collaboration may yield additional insight and data that can be used to improve the models in ORCA. There is a large biomechanical community including members within HRED and NASA and in sports medicine and orthopedics that could potentially be leveraged with respect to stress and trauma.

BRL-CAD

The presentation of the BRL-CAD effort was a model in terms of technical depth, external engagement with the broader scientific community, and articulation of the essence of BRL-CAD with a clarity that finally dispelled the uncertainties that many of the panel members have had for years.

The BRL-CAD program overall is a 20-year success story, which has grown into a tool that is carefully tailored to the needs of many of SLAD's analyses. The current plan has chosen its targets well. On the one hand, the move to incorporate non-uniform rational B-spline surfaces (NURB)-based boundary representation solid models will place BRL-CAD in the mainstream of modern CAD practice. It will greatly facilitate the development of a STEP translator,[6] which in turn allows much easier access to CAD models from platform vendors, eliminating one major bottleneck in the analysis process. On the other hand, enhancements required to support MUVES-3, especially with regard to moving parts and dynamic geometry, are necessary if that program is to reach its goals.

The technical approaches proposed are at or beyond the current state of the art. For example, BRL-CAD will implement a new and innovative surface-surface intersection algorithm from the literature, which promises the accuracy needed for building water-tight solid models for ray tracing. If successful, it will be a leapfrog technology.

The program is the best model that the Board has seen at SLAD of two-way interaction with the external technical environment. It actively participates in the open-source community and is used and contributed to by first-rate academic researchers. The program sports an excellent Web site with a Wiki, as well as a good Wikipedia article (an idea that other programs could adopt as well). It seems clear that BRL-CAD is of interest to the graphics and CAD research communities, that it rates very highly in terms of performance, and that it has gained wide acceptance by a worldwide community.

BRL-CAD uses a good mix between "making" where SLAD can add value (e.g., surface-surface intersection) and "buying" where it cannot (e.g., using an off-the-shelf STEP file parser from the National Institute of Standards and Technology), which acts as an effective force multiplier. However, some of its goals are technically challenging. For example, the revolutionary new surface-surface intersector may not work as advertised, or BRL-CAD's ideas for geometry comparison algorithms to support versioning of dynamic geometry may not pan out. These have been recognized as hard problems in the CAD community for many years. Risk mitigation plans are essential going forward.

One technical issue arose at the interface between MUVES-3 and BRL-CAD that the BRL-CAD group should pursue. The MUVES-3 presentation mentioned that the BRL-CAD ray tracing runs better

[6] A number of different file formats can be used to transfer data between mechanical CAD systems (e.g., STEP, IGES, DXF, etc.).

when single-threaded. This was such a peculiar result that the BRL-CAD group should be interested, since ray tracing is embarrassingly parallel.

Another area that demands careful attention to risk is the open-source software approach. The open-source project is a great innovation because the software is tested by a huge community and because contributions can come from a large number of developers. The open-source approach is one that has helped many small startup companies; these companies make their money on managed services built on their free software, which gets continual improvements from a spectrum of developers.

Open source does bring some risk, of course. It is imperative to have a security review of all software updates generated by the world external to ARL and to have acceptance criteria for the BRL-CAD application itself before using it on government computer networks. Assuming that ARL has constant monitoring of its networks, the BRL-CAD developers should develop monitoring rules for intrusion or exfiltration detection systems that are specific to BRL-CAD, such as well-known ports and message format validation. Further assuming that ARL has rules for using applications on classified networks, those rules should be reviewed specifically with respect to BRL-CAD.

OPPORTUNITIES AND CHALLENGES

MUVES-3

The Board was astonished, during its 2007 meeting at White Sands Missile Range, to learn that the initial operational capability for MUVES-3, SLAD's primary integrative software environment and interface between its many engineering models, had been deferred by several years (leaving the existing MUVES-S2 as the primary production tool for at least 5 years), and that the defined scope for the project had been significantly redesigned. The Board viewed this as a classic software disaster and recommended in the strongest possible terms that SLAD management undertake a detailed project audit and either kill the program or replan it at the earliest opportunity.

The Board did note that the correct metric had been identified for the success of this project, namely, the overall flow-time of an entire analysis project. Since this flow-time is typically dominated by setup costs, it is appropriate for the project to concentrate on streamlining and assisting setup rather than on minimizing execution time for the actual analysis.

It was evident that SLAD management took the Board's concerns to heart, making both organizational and programmatic changes to focus on the issues evident in the MUVES-3 effort. As a result, the project is in much better shape as the Board concludes this evaluation cycle, but the chasm still looms. The initial plan had a number of obvious risks. The programming language was not one that the developers were familiar with, the distributed architecture was new to the group, the existing system had to be maintained and enhanced while the new one was developed, there were few clear intermediate goals, and the overall objectives were somewhat murky.

A possible interpretation of what had transpired might be described as a fundamental underestimation of the amount of necessary knowledge and effort. This caused the managers to underestimate the amount of project management effort that was necessary. As the risks resulted in frustrating performance problems and schedule impacts, managers could not see what was wrong. Ultimately, the team learned the programming environment, brought specialized system experts to the forefront, corrected their mistakes by using more appropriate software packages, put better project oversight processes into place, and brought the skeleton prototype up to an acceptable level of performance.

The major performance problems seem to have been addressed in the current architecture. The benchmark data provide the validation of the performance goals, and the team's confidence in the architecture seems well founded. However, it is disconcerting to hear that the performance has been deemed good enough (with no particular justification) and that the most basic measure of performance comparison of MUVES-S2 to MUVES-3 (single-processor problem-solving speed) is buried in more complicated scenarios. The Board understands that MUVES-S2 lacked the important scalability property of being able to distribute the processing over multiple central processing units and multiple machines, but that does not mean that the basic speed comparison is entirely irrelevant.

The parts that have been changed seem to be well-known risk factors for distributed architectures and/or high-performance systems, and the development team and its management should take time to understand why the risks were underestimated and why it took so long to address them. It may be necessary to make changes in parts of the architecture that have yet to be realized, so it is important that the team be prepared to deal with problems quickly and effectively.

One problem might have been that the team relied on many Java services that are attractive in their power and ease of use without understanding their performance aspects. Because there was (apparently) no detailed breakdown of the design into implied performance requirements (such as the time for a remote procedure invocation), it was not possible to determine if a service would be suitable for MUVES-3 without building a prototype system. A detailed performance model would have allowed developers simply to measure the remote procedure invocation overhead and have an immediate decision about suitability. So, it seems that the team had to build one prototype system after another, searching for a mix of Java packages that finally meshed. In a positive sense it can be said that everyone probably learned a lot, but this is a far cry from the way that R&D works in industry.

The architecture now relies on two important pieces of open-source software for the Java environment: Rio (for the distributed processing) and Java Spaces (for the run-time data). Overall, it is good to see SLAD taking advantage of the strengths of the large open-source projects and that it is a project contributor (to Rio). This must be done with careful thought as to the certification of the resulting product for use in sensitive or classified environments and to the security of ARL computing resources.

The Board was not provided any supporting data about how the design changes, such as using the peer-to-peer system or batching the BRL-CAD requests, improved performance over the master-worker scheme. The Board heard in passing that BRL-CAD runs better when single-threaded; that seemed interesting and worth investigating. Is there anything in the Iteration 3 investigations that yields insight into how to allocate resources for analyses that test the performance boundaries with respect to resources? Can a systems expert help an analyst determine how many machines of what type are necessary for large problems? Are there internal parameters that can be tweaked by an expert to improve the run-time of an analysis (note, the tweaking might be done at the inception of the analysis; the developers seemed to feel that changes to parameters could not be made during run-time, but the issue of pre-run-time configuration tuning was not addressed)?

The demonstration left an uncertain definition of what the Geometry Service in the architectural diagrams entails. The notes on the presentation slides say that it is the ray-tracing engine (BRL-CAD). However, the Geometry Service was later described by SLAD as "to be determined." When the panel asked how the demonstration was done absent the Geometry Service, the answer was that it just read a file. If the panel saw a demonstration that did not actually invoke the BRL-CAD ray-tracer, the panel should have been informed up-front. The panel members remain unsure on this question and think that the Geometry Service might simply be a service that takes an object name and finds its geometric description in a database.

There are other ways to approach the architecture issues, and it was not made clear by SLAD why distributed computing by way of commodity servers and workstations was selected as the computing base. Was that a deliberate decision or happenstance? High-performance scientific databases for scientific computation run quite well on cluster machines, for example. Did the team do a cost-benefit analysis of the hardware base? Would SLAD have been better served by simpler software on more expensive machines? Although it was encouraging to hear that the developers consider their architecture stable (again), it would have been helpful to provide a better understanding of the risks involved if it is necessary to tweak things (again).

The Board suspects that SLAD does not have much current experience developing complex software systems. BRL-CAD was developed partly as the result of one dynamic and charismatic individual, and it has been a good foundation for further development. MUVES seems to lack a similar kind of star technical leader.

There are many ways to approach the project management problem for a system like MUVES-3, and SLAD could have tried a small tiger-team approach for 9 months as a possible alternative. Or, if it had clear performance and functional criteria, it could have contracted out the initial architecture design and prototyping. Perhaps the design team was too fragmented initially, working part time on MUVES-S2, part time on MUVES-3; if so, the organizational unification of the two teams should help. Perhaps the problem was inherently more complicated than originally realized—could more frequent reviews and benchmarks have delineated the complications earlier?

Some years ago the project manager of MUVES told the Board that Java was the language of choice in large part because recent college graduates had a strong preference for Java over C or C++. The Board wonders if the commitment to Java resulted in the hiring and retaining of recent graduates with computer science degrees.

Does MUVES-3 have any need for security features such as authentication, access control, and encrypted data (either in a repository or on-the-wire)? SLAD reported that authentication was being considered, but are there elucidated security requirements that can be mapped to an architecture? It is encouraging that the management is working with a software development risk-mitigation consultant and there is hope that this work leads to a deeper understanding of how to control the development cycle for complex software projects in the future. The Management Review Board seems like a good idea, but it remains to be seen if it can handle the fundamental design problems that still lie ahead.

An important question is whether SLAD learned any lessons that give it confidence that it can approach similarly complex software systems in the future. Correspondingly, SLAD should document the lessons learned that can inform future significant efforts. S4, SLAD's other major software development effort, is likely to provide many further opportunities to apply the lessons learned in the MUVES-3 project. The methods and results of the MUVES-3 performance analysis and design decisions would probably make an interesting paper for a distribution systems applied technology conference. Could the developers set a goal of producing at least one paper for publication each year?

Finally, the future is not clear with regard to the performance improvement in overall analysis time that currently motivates the MUVES-3 effort. This subject was addressed in the 2008 MUVES-3 presentation to the panel in which MUVES-S2 data were presented, indicating that less than 15 percent of the total analysis process time (measured as effort in person-weeks) was represented by run-time and deliverables.[7] The average percentage of preparation effort and the MUVES-3 features aimed at reduction in preparation time for five process phases in seven types of analysis were identified (slide 82), as shown in Table 5.1.

No attempt has been made in the figures shown in Table 5.1 to weight the percentages by the importance or relative numbers of analysis types, which range from air vehicle survivability to spare

[7] Mark Burdeshaw, Army Research Laboratory, "MUVES 3 Overview," presentation to the Army Research Laboratory Technical Assessment Board, Aberdeen Proving Ground, Maryland, July 23, 2008, Slides 78 to 90.

TABLE 5.1 Average MUVES-S2 Preparation Effort, by Process Phase, and MUVES-3 Features Aimed at Reduction in Preparation Time

Process	% Time	MUVES-3 Feature
Target description	24.4	BRL-CAD Geometry Service
Criticality analysis	22.4	GUI: Fault Tree Editor
Behind-armor debris	2.4	Collaborative Environment
Input data	13.4	Collaborative Environment
Pre-run review	22.2	Collaborative Environment, MRB, Test
Total preparation	84.9	

SOURCE: Based on Mark Burdeshaw, Army Research Laboratory, "MUVES 3 Overview," presentation to the Army Research Laboratory Technical Assessment Board, Aberdeen Proving Ground, Maryland, July 23, 2008, Slides 78 to 90.

parts analysis. It is clear, however, that preparation for the run requires the major fraction of effort in the MUVES-S2 system.

What is not clear in SLAD's presentation is the nature of the corrective action to be incorporated in the MUVES-3 program. Except for the improvement in providing a more user-friendly BRL-CAD input system, steps to be taken and the economies to be realized in the MUVES-3 preparation process remain unclear. The Fault Tree Editor and Collaborative Environment have not been defined to the point where a huge reduction in workload can be predicted with confidence.

Active Protection Systems

The active protection systems (APSs) will become a critically important element of ground platform survivability in the coming years. In fact, an argument can be made that the U.S. Army is late in this area. The threat environment of future battlefields is expected to be even more complex and more lethal than it is today. The ability of Army acquisition programs to evaluate operational requirements relating to APSs and to assess candidate system performance and suitability is an important aspect of fielding APS capabilities. SLAD has the expertise to enhance APS modeling and simulation capabilities in several key areas.

SLAD personnel presented a methodological framework for evaluating APS protection of ground vehicles. SLAD's recognizing that APS modeling needs to be approached in an end-to-end fashion instead of piecemeal is commendable. SLAD's using the existing Army Research, Development, and Engineering Command (RDECOM) APS model instead of building a new model is a cost-effective approach that will provide needed modeling and simulation capabilities sooner is an appropriate choice. The Board was presented a threat analysis, including the operational aspects of multiple, simultaneous threats and threat signature modeling. SLAD also reviewed work being done to model threat warning systems with respect to evaluating potential false alarms, an analysis of threat characteristics versus other battlefield signature sources, and a possible approach to discriminating threats from battlefield clutter.

However, the Board was not shown a system-level analysis. So, whereas SLAD is making incremental augmentation of the RDECOM model with improved representation of phenomenology, the Board was not able to assess whether all of the critical elements affecting APS performance have been identified and if they were being adequately addressed. In fact, there seems to be at least one critical aspect of APS performance that is not being addressed—the overall time line associated with the threat engagement sequence. The HRED IMPRINT model may be of use in estimating task times in this sequence. APS performance will largely be determined by its ability to launch active countermeasures in a timely manner while maintaining a very low false-alarm rate. SLAD should focus on the APS response time line issue.

SLAD appears to be behind the leading edge in mid-infrared threat warning sensor phenomenology. This may be an area where SLAD could benefit from collaboration with other government organizations and industry. There is considerable work being done on airborne missile warning systems that should be directly applicable. In addition, the lack of actual algorithms used by prime contractors represents a major obstacle to SLAD's focusing efforts in the highest-impact areas.

The APS project appears to be a candidate for application of the SoS concept, which would allow the individual elements of the threat-sensor-response process to be addressed in a comprehensive way, bringing in the depth of SLAD knowledge and modeling of the battlefield environment. The APS would provide an example of how this approach, presented by SLAD as a rather abstract concept, could be brought to bear on a critical problem that until now has been attacked piecemeal by various contractors and RDECOM, with only peripheral participation by ARL.

Mission and Means Framework

The Mission and Means Framework has been applied successfully in a number of different instances since its inception. Although the panel members are not unanimous in their support of MMF as a methodological advance, use within the SLAD customer set is a positive indicator. However, SLAD does not appear to be devoting additional effort toward developing the methodology and addressing previously noted technical shortfalls such as an explicit approach to accommodating stochasticity. Indeed, the applications of MMF shown to the Board were predominantly performed by Dynamics Research Corporation (DRC), an independent defense contractor. The use of DRC for production while SLAD continued development of methodology would be a defensible approach, given limited human resources. However, SLAD's abandonment of an immature methodology to an independent contractor holds reputation risks for ARL. If the contractor applied the methodology inappropriately, would it take responsibility for errors, or would it note that it was just deploying an SLAD-developed tool? In any case, if SLAD discontinues MMF development, it will then be completely dependent on a single tool, S4, for SoS methodology.

OVERALL TECHNICAL QUALITY OF THE WORK

As noted previously, the work performed by SLAD is technically competent, especially in the context of rapid response to operational needs where time and resources are severely constrained. The SLAD staff has extensive experience with this type of analysis and possesses great domain knowledge concerning specialized systems; they are qualified for the work.

In terms of significant, multiyear efforts requiring significant development of models, tools, and analysis methodologies, the verdict is mixed. BRL-CAD is proof that SLAD is capable of significant, even state-of-the-art software development embodying complicated technical underpinnings. MUVES-S2 demonstrates that SLAD is capable of integrating many disparate tools to create a metatool of fairly broad applicability. However, the principal lesson of the MUVES-3 project is that SLAD failed to develop a management construct, including requirements definition, milestones, reviews, and diagnostics, up to the task of modernizing its existing software environment. This shortfall in systems engineering is far from unique in the national security technical community, and it is encouraging that SLAD management is taking the lessons learned seriously.

Another area of concern to the Board is whether the extremely granular nature of the SLAD portfolio, compounded by the many rapid-response tasks generated in the course of the war effort, will *ever* permit SLAD staff to develop the skills necessary to perform larger-scale, longer-term development or analysis efforts. This is particularly bothersome in the context of the Army's need for SoS analysis methodology for network-centric operations on an irregular battlefield. SLAD is currently depending primarily on a single team of academic collaborators at New Mexico State University to provide the tool set needed for that transformational and essential task. This should be matter of significant concern to ARL management and Army leadership.

In contrast, the Board has been frustrated for years about the insularity of SLAD staff with regard to the larger scientific community and the reticence of SLAD to professionally engage with that community. SLAD staff are clearly frustrated with this message, and to the extent that ARL resource management serves as a constraint (in providing travel and conference funds), justifiably so. However, for SLAD staff to develop professionally as their counterparts in other federal, university, and industrial laboratories do nationally and internationally, there is really no alternative to more extensive and prolonged professional interaction.

SLAD has new mission and vision statements, notably including the aspiration for the SLAD staff to be unsurpassed in dedication and willing to do whatever it takes to support the warfighters. The Board has long been impressed by the dedication of SLAD staff in supporting our nation's warriors, especially when measured in terms of long hours, personal risk, and making do with available resources. The Board exhorts SLAD to be similarly dedicated and willing to do whatever it takes to better engage the external scientific community to leverage its findings and results in furthering the SLAD mission. The lesson of the excellent results obtained through SLAD partnership in JTAPIC should help motivate a much broader engagement than currently exists (significant counterexamples such as BRL-CAD notwithstanding).

That the SLAD portfolio does not lend itself as readily as those of the other directorates within ARL to external collaboration, publication, and conference participation should not serve as an excuse to an appropriately motivated staff. SLAD insularity significantly compromises the directorate's ability to leverage academic and commercial developments, especially in such areas as computer and network security, biomechanics, and software development, where investment outside the Army dwarfs organic resources and capabilities. Academic collaboration is also a key to strategic workforce development, since the exposure of graduate and undergraduate students to highly relevant applied research and development may enhance SLAD's recruiting pool. SLAD staff has shown increasing involvement in conferences and professional societies in recent years. Funding constraints and demands for support of current military operations appear to have recently blunted this improving trend over the current assessment period. SLAD and ARL management should resist the temptation to allow current short-term pressures to cause a relaxation into a more insular posture. Strategic workforce development, as well as longer-term Army needs, demand that SLAD staff seek professional enrichment and involvement in the broader technical community.

6

Vehicle Technology Directorate

INTRODUCTION

The Vehicle Technology Directorate (VTD) was reviewed by the Panel on Air and Ground Vehicle Technology of the Army Research Laboratory Technical Assessment Board (ARLTAB). The directorate has three divisions (Mechanics, Propulsion, and Unmanned Vehicle Technologies) and one program office (for management of the Army Research Laboratory [ARL] Robotics Collaborative Technology Alliance [CTA]) that were reviewed by the panel.

Appendix A shows the funding and staffing profiles for VTD (see Tables A.1 and A.2). The assessment below reflects visits by the Panel on Air and Ground Vehicle Technology to the VTD sites at the NASA Glenn Research Center (August 15-17, 2007) and the ARL facilities at Aberdeen Proving Ground, Maryland (June 2-4, 2008).

CHANGES SINCE THE PREVIOUS REVIEW

Many significant changes have occurred since the 2005-2006 review of the Vehicle Technology Directorate. VTD began relocation to Aberdeen Proving Ground, Maryland, as part of the 2005 Base Realignment and Closure (BRAC) action. This move resulted in major adjustments in every aspect of the operation of the directorate, especially for the Propulsion Division. The move presents an opportunity to centralize the activities of the directorate and to align activities more closely with Army needs and other Army organizations. The ARL and VTD leadership is effectively moving forward in the face of these major events. In addition to the fiscal and facilities changes, planned program changes include reduced activities in some technologies (e.g., active rotor technology development and large-turbine-engine concepts) and increased (or refocused) activities in other technologies (e.g., microsystem mechanics, small-engine technology, and prognostics and diagnostics). To some extent, these changes also reflect changes in facilities and access to laboratories and equipment previously shared with NASA. The Board exhorts the Army to support ARL's efforts to maintain effective levels of staffing and equipment in order to continue essential work in such areas as propulsion and aircraft structures and materials.

ACCOMPLISHMENTS AND ADVANCEMENTS

Significant accomplishments were achieved by several divisions and programs of the Vehicle Technology Directorate during the past 2 years. Several parts of the propulsion effort, which is potentially at some risk in the transition from NASA Glenn, showed advances. The Active Stall Control Engine Demonstration (ASCED) (also noted in ARLTAB's 2005-2006 report[1]) continues to combine state-of-the-art analysis with full-engine tests to advance the understanding and control of performance changes during service in environments of interest to Army missions. Other programs that have shown promise include the development of technology for high-efficiency wave-rotor-topped gas turbine engines, with initial results showing reduction of the specific fuel consumption by about 15 percent while increasing the power-to-weight-flow ratio by about 18 percent. In the context of increasing cost (and threat of loss of supply) of battlefield fuel, work of this type has obvious importance in reducing gas turbine engine specific fuel consumption while increasing the power-to-weight flow—progress that will make a significant contribution to Army missions. However, this work, which began as a NASA/ARL effort in the 1990s, has a rather long technical horizon for application, and it is also subject to possible disruption by the BRAC activities. In general, the engine research is appropriately focused on a balanced program of near-term and fundamental research. The computational fluid dynamics (CFD) simulation of compressor stall avoidance and the work on hot restart is excellent work that addresses near-term operational issues and provides a sound foundation for future developments and more refined research.

A second example of a significant advancement is the Robotics Collaborative Technology Alliance. The Robotics CTA is a well-organized and well-executed interlocking consortium of industry, academia, and government laboratory personnel that seems to offer a best-practice model for VTD and ARL. It was established through a competitive pre-award process and is managed in a centralized but intellectually fluid process capable of adapting to the changing features of the research landscape in this still-maturing field. The Robotics CTA presentations evidenced state-of-the-art and often pioneering results from some of the most qualified researchers in their fields. An example of cutting-edge research is the real-time extraction of geometric and semantic terrain representation from raw ladar point clouds. An example of the intellectual fluidity in allocating new resources to track potentially game-changing advances in technology is provided by the new RIVET simulation environment. A growing number of transition successes into fielded application platforms within the Tank-Automotive Research, Development, and Engineering Center (TARDEC) and the Future Combat Systems (FCS) validate the up-front positive impression conveyed by the CTA program portfolio itself. The very high quality research and its record of well-knit practical integration in Army-relevant field demonstrations at the Fort Indiantown Gap, Pennsylvania, facility suggest that this approximately $10 million per annum investment—roughly one-third of the VTD budget—is paying off.

Finally, VTD is moving toward building its effort in the health and usage monitoring (HUMS)/condition-based maintenance (CBM) field with the addition of the new Mechanics Division chief. This should be encouraged as a real opportunity to apply the existing VTD expertise in rotorcraft, composite materials, fracture/fatigue, and nondestructive evaluation and diagnostics into an area that has a strong potential benefit to the Army. Structural health monitoring, HUMS, CBM, and so on constitute a rapidly

[1] National Research Council, *2005-2006 Assessment of the Army Research Laboratory,* Washington, D.C.: The National Academies Press, 2007.

growing field that requires the integration of sensors, signal processing, mechanics, and material behavior. VTD's effort to achieve excellence and critical mass is a significant program advance, which could be significantly strengthened by putting together a collaboration similar to a CTA that might include the rotorcraft industry, sensor companies, and university researchers. Close integration with other activities in this field is also encouraged.

OPPORTUNITIES AND CHALLENGES

At this time of realignment and redefinition, it is especially important to maintain a systems focus, such that each individual researcher is able to clearly state how his or her research, if successful, will enable additional desirable capability for the warfighter.

VTD has a window of opportunity because of the changes dictated by the BRAC to ensure that all of its programs across the VTD divisions mutually support one another and that all programs are aligned to meet the needs of the Army. For example, the Robotics CTA is an excellent program that is clearly demonstrating an approach that is producing a great leveraging of ARL's limited funds and personnel to produce the artificial intelligence and vision necessary for a robot to autonomously get from point A to point B. However, there were no presentations from the Propulsion Division or the Mechanics Division indicating that they were developing the supporting technologies in their areas that would be needed by these robots. The directorate should use this window of opportunity to ensure that it has an integrated program across all of its divisions. In addition, the CTA approach, which is demonstrating excellence, should be considered in other areas, as appropriate, to leverage VTD limited personnel and funds to produce the technology needs of the Army.

VTD is in the process of establishing a group that will have responsibility for integrating the portfolio of research and communicating both internally and externally. The establishment of this group is appropriate. It should be responsible for items such as the following:

1. A clear statement of the directorate's vision, related to Army needs;
2. A statement for each division that defines how its portfolio of research in total meets the directorate's vision and mutually supports other divisions;
3. Notional definitions of vehicles of each type required by the warfighter, to focus the directorate's vision and to ensure that key technologies are not missed;
4. For each program, limit calculations that show how much of the total potential capability would be enabled by a successful completion of the research project(s), to help to focus the researcher on the importance of his or her research; and
5. Identification of crosscutting technologies and disruptive concepts and technologies for shared responsibilities and focus.

The directorate is undergoing changes as it consolidates its workforce at Aberdeen Proving Ground. In particular, there is new staff in the Mechanics Division; the Board looks forward to this staff's establishing a portfolio of research programs that meets the directorate's vision. Similarly, the Propulsion Division is in large part moving from NASA Glenn. The Board recognizes several improvements in the Propulsion Division's research portfolio and looks forward to its continued development.

VTD is rigorously involved in a strategic planning activity that will bring the entire structure of VTD and its divisions into focus. As enumerated below, at least three areas of crosscutting issues are appropriate for discussion during that planning effort.

One of these crosscutting needs is to define a platform for the future that will identify, for example, what the next helicopter or ground vehicle engine, robotic system, or other system is going to be (to the extent that is possible), what its goals will be, and what technologies are necessary to achieve these goals. Progress in this effort will help to set consistent, shared directions in a directorate that is clearly addressing future needs in addition to essential present improvements.

A second opportunity has to do with awareness of the technical activities and horizons in the community at large. This is especially challenging, considering the remarkable technical scope of VTD. VTD should continue to emphasize refereed publication of advances and the participation of investigators in teams, partnerships, and cooperative activities with other organizations. Areas of particular importance include analysis and computation (e.g., predictive methods for material properties from first principles) and systems analysis. A companion issue is the question of focusing on a comparatively small number of fundamental issues of broad importance versus the expedient (but sometimes isolated) addressing of many technical matters of smaller scope. The greater community's awareness and perception of directorate technical activities and leadership are important in the ability to leverage the work of others and to recruit and retain the best individuals for Army laboratories. Being seen as the place to go to work on state-of-the-art technologies is a worthy goal, deserving of an investment of time and resources to ensure achievement.

A third opportunity is the consideration of shared capabilities and facilities for computational work. Analysis and computation are becoming (with good reason) a more consistent aspect of what the directorate (and everyone else) does. There is a special opportunity for VTD to generalize its capability in this important area and at the same time to focus on a few areas; to support and create data sets, especially those specific to VTD experience; to validate codes and establish diagnostics; to organize round-robins; and to interpret results. Given the Army-specific advantage of data sets for many specialized hardware embodiments, this is thought to be a significant opportunity for leadership. The success of shared objectives turns on communication both within the directorate and with the greater community of investigators at large whose work and insights can be leveraged.

OVERALL TECHNICAL QUALITY OF THE WORK

The Vehicle Technology Directorate has established the tradition of a research approach that has successfully applied analytical tools and experimental methods in controlled environments to hardware-based problems at various scales. With new directions and realignments under way, it is especially important to revisit the need for a statement of specific requirements, goals, and schedules for each individual project. Exploratory areas (such as flapping wing structures and self-healing) are certainly appropriate for best-effort work for an initial trial period, but long-term goals and deliverables in the Army context are final requirements. Bringing new technical horizons into the mix (e.g., robotics and unmanned vehicle technologies) presents new opportunities and challenges to the task of establishing methodology. For example, while the organizational research model is to be applauded and the notable per-project success rate within the Robotics CTA is to be recognized, there are a number of broader issues that VTD and ARL might wish to consider as this and similar activities move forward. Foremost, despite the growing number of single-point successes in transitioning CTA technology to more-applied Army programs, it is not clear that individual projects' principal investigators, or even the CTA central leadership, have been able to find a fundamental, long-term view of the CTA's mission within the Army. This may reflect the constraints imposed by the Army's continued focus on FCS as its defining activity center for robotics.

One programmatic consequence of this perceived standoff from direct soldier-on-the-ground problem statements may be the seemingly incomplete vision of how the various constituent perceptual and reasoning capabilities now being vigorously developed, fielded, and transitioned will be integrated and deployed in functioning Army systems. A corresponding intellectual feature of this standoff is the very premise that perception and intelligence capabilities may be split off from equally fundamental considerations about the mechanical systems and their environmental settings. VTD should develop specific requirements, goals, and schedules for each individual project, reflecting systems engineering analyses that clarify the links of the projects to Army needs.

The organization of VTD's new Unmanned Vehicle Technologies Division provides a useful opportunity to reassess this premise and to explore the extent to which intelligent bodies and minds are linked by the environments within which they carry out specific mission capabilities. On a more general level, bringing operational and human-factor objectives into the methodologies that enable research success without diluting the rigor of the fundamental research is a paramount challenge, but worthy of address. As it happens, VTD and ARL have many of the elements of that discussion at hand, with a strong foundation in mechanical systems and a growing excellence in intelligent autonomous systems. This is an outstanding technical environment in which to make those associations and connections.

A benefit (or at least opportunity for benefit) of the BRAC activities is an enhancement of the VTD contributions to the Army's needs, although the record of VTD in this regard is already generally excellent. This is especially true in the traditional areas of materials and propulsion, and it is increasingly true in the new technical directions of robotics and unmanned vehicles. Near-term benefits from work on engine and gear box deterioration (e.g., ASCED) and work on materials degradation and damage detection (e.g., Air Coupled Thermography Inspection) are easy to identify. Propulsion and critical structure research and development translate directly into extended equipment deployment, extended missions, and reduced demands on depot assets.

Other work may have longer lead times but great potential effect. The directorate has a long history of excellence in high-temperature materials that have the collective capability to create game-changing capabilities in warfighter vehicles. An example of this type of effort is the ceramic composite and coatings work, which has the additional advantage of industry partners. The development of unmanned engines should also be mentioned in this context. And the development of analytical and computational methodologies and capabilities is a clear investment in future design and development capabilities, especially for active rotor design, tiltrotor aeroelasticity, high-resolution CFD, and a host of nonlinear problems associated with technologies such as flapping wings. Examples of this work include the Parallel Unsteady Domain Information Transfer effort and the development of robot algorithms for uncertain environments. Propulsion for unmanned vehicles would also appear to be an opportunity. At a more general level, while the Army Energy Program addresses installations and many Army programs address soldier power, vehicle power and energy would appear to have a natural home in this directorate.

It is clear that the VTD programs are contributing to the greater technical community at the fundamental and applied levels. Much of the high-temperature material work being done by VTD personnel, especially in cooperation with NASA Glenn Research Center, is unique and essential and is not being emphasized by many (if not most) other mission organizations or by academia. Some elements of the rotorcraft work are also clearly on the forefront of technical community efforts. Efforts to maintain awareness and involvement in frontier work at the community level need to be redoubled in some cases.

As an example of this need, the active-passive rotor performance project is a refocused effort from prior smart rotor (active twist) activities in the noise and vibration area to assess performance. This refocus is based on this review's (and other) comments indicating that improved performance is the key attribute that needs to be proven to justify active rotor applications. At the moment, the effort involves analysis only. The project staff is a talented group of investigators with expertise in this area, and they are focused on a topic of significance to the community—performance—but at this early stage, in some respects, they are playing catch-up to others in the field. This work would benefit from the identification of a path that will distinguish this effort from others in this area. One opportunity that does exist is to seek collaboration in the upcoming NASA Ames Research Center testing on Boeing and Sikorsky active rotors (which may be in advance of the next VTD active twist rotor test in late 2009) and to see if those data can be the catalyst to the VTD work.

A second example is the structural dynamics for rotorcraft activity, which is an effort to address one building-block component of comprehensive rotorcraft analysis—that being the dynamic modeling of redundant, nonlinear airframe structures. In reality, this is one element of a very broad and robust rotorcraft community of existing and past efforts along these lines. The focus on addressing fastener/bolted joints has been the subject of prior work (e.g., by the National Rotorcraft Technology Center and Rotorcraft Industry Technology Association). This effort will benefit from detailed discussions within the structural dynamics community to best define an approach that can leverage past work.

A third example has to do with mesoscale flapping wing structures. The scope of the VTD work, to design and construct mesoscale flapping wings capable of generating forces similar to those generated by a fruit fly, is impressive and applauded. The study combines experimental and modeling work on millimeter-scale flapping wings. The modeling so far is limited to two-dimensional modeling using corrections (history integrals, added mass) to the quasi-steady formulas for lift and drag. The Reynolds number is larger than 1 but still low so that viscous effects are significant. The Strouhal number for the flapping action is of order 1 so that the unsteady effects are significant. Because of the values of these similarity parameters, classical wing theory does not apply, as the investigators recognize. A CFD solution rather than the modeling with corrections should be pursued. Clear objectives and a systematic approach could result in a considerable contribution to the greater community, because this is a research area of very broad activity with support coming from a variety of agencies and organizations. Well-defined goals and specific concentrations will help to ensure success in this context.

7

Weapons and Materials Research Directorate

INTRODUCTION

The Army Research Laboratory's (ARL's) Weapons and Materials Research Directorate (WMRD) was reviewed by the Panel on Armor and Armaments of the Army Research Laboratory Technical Assessment Board (ARLTAB) at Aberdeen Proving Ground, Maryland, during June 5-7, 2007, and July 15-17, 2008. The theme of the 2007 review was materials research performed in the directorate; the 2008 review was related to research and development (R&D) performed in the lethality and survivability areas.

CHANGES SINCE THE PREVIOUS REVIEW

In prior years, the Board had commented that the Weapons and Materials Research Directorate had not appeared to be striking an appropriate balance between experiment and computational efforts, with too little emphasis on computational and modeling areas. However, during this assessment period the panel was briefed on the full scope of WMRD's programs, and it appears that the balance has improved considerably. A new effort, systems effective modeling (SEM), reviewed for the first time, holds promise. SEM is intended to provide ARL-WMRD with a systematic approach for evaluating the usefulness of existing and new weapons systems during design and development stages. As an example, SEM was applied to the development of a 7.62 mm lead-free round, and its application is accelerating the deployment of lead-free rounds that will provide the same or improved lethality. In its general application, SEM promotes the development of simulation and modeling tools and a series of experimental studies to test the validity of these tools. In an iterative process, the simulation and modeling tools are refined to the point that designers have confidence in the predictions resulting from these models. This is a crucial set of steps, which all modeling efforts should seek to emulate. SEM is an excellent tool that provides a methodology through which to incorporate modeling and simulation productively and thereby to accelerate and reduce the costs associated with design and deployment. It is an essential capability that should be expanded to other ARL programs. As an example, SEM might have a profound impact on the programs involving quantum chemical modeling by combining modeling efforts with the problem of identifying experimental programs that will serve to validate and instill confidence in the models.

Also newly presented was the development of the Novel Energetic Research Facility (NERF), which led to the development of DEMN, an explosive fill. This effort demonstrates a continued paradigm shift from looking for one magic explosive material to using calculations and experience to suggest a mixture of materials providing tailor-made performance.

During the 2006 review, the panel learned of a new program—Soft Tissue Physics and Applications—and offered numerous suggestions for its improvement. During the 2008 review, it was obvious that WMRD had taken those suggestions to heart, and the group is to be commended for its successful effort in refocusing the program, which addresses a national defense priority and now seems to be going in the right direction. The understanding of munitions-induced trauma is of high importance in the design of armor, in the design of armaments (e.g., to reduce the risk of collateral damage to noncombatants), in the design and delivery of effective treatment of the wounded, and in the training of medical personnel. Blunt trauma and traumatic brain injury (TBI) have very high profiles with the public and the Congress; hence, WMRD's efforts in modeling blast loading are timely.

ACCOMPLISHMENTS AND ADVANCEMENTS

Materials

A very important element of the materials effort in WMRD is the formation of several Materials Centers of Excellence (MCoEs) in which the scientific input of academia is melded with the technology-driven, warfighter-focused programs at ARL. Currently, five MCoEs are being funded—at the Johns Hopkins University, Rutgers University, the University of Delaware, Virginia Polytechnic Institute and State University (Virginia Tech), and Drexel University. Some of their contributions include the following: (1) work on developing phase diagrams for glassy grain boundary phases in grain boundary engineered boron carbide—necessary for improving the consistency of armor protection of this very lightweight ceramic armor; (2) development of magnesium (Mg) as a lightweight metallic armor; (3) nanocrystalline tungsten (W) to replace depleted uranium (DU); and (4) nanocrystalline aluminum having the strength of steel but one-third the density. In addition to their scientific input, the MCoEs have been the source of a number of summer postdoctoral researchers, some of whom have stayed on as staff members.

Metals

The Microscale Compressive Properties of Metallic Glasses project, undertaken in collaboration with the Johns Hopkins University MCoE, explores new microcompression techniques as a method of assessing the properties of metallic glasses. The goal is to measure physical properties of microconstituents to compare with bulk measurements and to inform the models of metallic glasses and composites in the future. This tie to modeling work was not highlighted by WMRD, but it seems to be the only practical use for the results of this work. Extreme care needs to be taken so that experimental variations (notably alignment) do not dominate the observed results. WMRD is encouraged to leverage these measurements with its multiscale modeling interests, because only through linking the micromechanics to larger length scales can it be hoped that the research described will provide new engineering tools of relevance to Army needs. As this technique continues to be developed and tied to single-crystal and multiscale constitutive modeling, in particular at Wright-Patterson Air Force Base, WMRD should seek its own unique niche in the leveraging of this technique to underpin improved science and technology needed to address Army needs.

Composites

Composite materials continue to play an important role in the Army mission in areas of lethality and survivability. Among the various materials-related research activities within WMRD, those focused on composites appear to be among the strongest. The work on transparent composites, which are needed for use in both combat and tactical vehicles, is particularly impressive. Faceplates, vehicular armor, and different grades of ballistically resistant composites were described, but some discussion regarding the scratch resistance of these transparent materials would have been useful. A significant amount of work has been performed on spinels with Technology Assessment and Transfer, Inc., a Small Business Innovation Research (SBIR) contractor. Large plates about 12 inches square were being produced, and a Department of Defense (DoD) Manufacturing Technology/Cooperative Research and Development Agreement (CRADA) program is being pursued to expand these dimensions to handle windshields and other applications for both the Army and the Air Force.

Most current projects are strongly driven by and responsive to the near-term needs of the Army. All appear to have good potential for significant payoff. Many of these projects are Army-specific and would not likely be addressed within other organizations, reaffirming the need to maintain internal expertise in the area. The corresponding MCoE at the University of Delaware is well integrated with the ARL effort and continues to make strong contributions. The composites activity has benefited from a solid group of young, bright, energetic individuals. Evidently, recruitment efforts have met with good success. These efforts should be continued. In addition, in order to retain these individuals and maintain internal expertise in the long run, management should take an active role to ensure that its scientists reach their full intellectual potential. This includes encouraging them to publish the more fundamental work in peer-reviewed journals and providing them with the necessary time and resources; promoting interaction with the broader scientific community through participation in conferences and workshops; and helping them establish collaborations with recognized experts in their fields of work, not only with the MCoE but also other institutions.

Transparent Composites The work here emphasized the need to create transparent composites for use in both combat and tactical vehicles. The efforts ranged from making transparent polyurethanes to introducing nanoporous polymers in glass matrices to achieve transparency. The work described was focused toward an immediate Army need and was more in the 6.2 than the 6.1 category. Most of the work described involved empirical experimental approaches based on pure materials selection principles and did not involve any predictive modeling. Residual stress was shown to play a major role in the transparency of the product produced. Approaches using chromophores were minimally successful and should be continued further, as there appeared to be some possibilities for success. A poster on this topic was very comprehensive and much more elaborate and addressed many of the difficult but solvable issues.

Electromagnetic Gun Rail Wear The problem of rail wear has been worked on for more than 20 years within the DoD. The novelty of the work described here is in the use of novel materials for substrates and the use of cold spray coatings to reduce the wear and erosion during firing. (Cold spray is a relatively new technique that has been developed within ARL in the past 2 years.) Efforts are being made to deposit tungsten coated with copper to improve the wear behavior. In addition, efforts are being planned with tantalum (Ta) and molybdenum (Mo) coated powders so that higher temperatures can be withstood (repeated shots) during firing. However, heat extraction rapidly between firings may become a rate (firing)-limiting step, since many of the materials do not exhibit the excellent thermal conductivity of copper. Additionally, elevated temperature exposure may also result in oxidation of the Ta and the Mo, resulting in wear debris that may be more catastrophic than the base materials. Thermal expansion issues are also very difficult to address in this application, and the rate of heating and cooling may dictate the overall life of the barrel and its effectiveness in target accuracy. This work is empirical and experimental in nature, which is probably justified because modeling these conditions, which are dynamic under firing, is difficult without certain specific start and end data points. Work in this area is of direct relevance to the Army's desires to produce the next electromagnetic (EM) gun. However, considering the numerous challenges involved, it would appear that this is in the category of a long-term program that may generate a significant amount of information that may eventually lead to the actual fielding of an EM gun. Coordination of this effort with the Navy and the Missile Defense Agency should be encouraged, so that a number of the lessons learned can be captured, documented, and retained for the newer generation of staff dealing with research on this topic.

Electromagnetic Rail Gun Composites The project on EM rail gun composites involves the design, materials selection, and construction methods employed in the construction of a prototype rail gun. This project highlighted a variety of complementary techniques used effectively in parallel, including finite element simulations of mechanics and electromagnetics, experimental measurements both in situ and ex situ during the firing of the gun, and empirical design of composites. The flavor of this project is clearly on the empirical side, but it uses state-of-the-art diagnostics and analysis. This project stands out as a prime example of materials technologies transitioning into applications of great relevance to the Army.

Ceramics

The focus of the WMRD work on ceramics is on trying to develop sintering methods for B_4C that are reproducible from batch to batch and also concurrently provide repeatable ballistic performance. Sintering of B_4C has not changed in more than 20 years, and much of the current product has been performing less than satisfactorily in the field. The recent discovery of shear banding and amorphization in B_4C led to an urgent quest to understand the behavior of these materials at very high strain rates. Grain boundary engineering approaches have been tried on metals for many years, and there has been some extensive modeling that has accompanied heat treatment of materials to improve hardness as well as to improve resistance to effects like hydrogen embrittlement.

ARL's approach is focused on creating a glassy intergranular phase that could wet the grain boundaries and assist in promoting preferred fracture pathways. The materials that are being tried are mostly those that form glassy phases, especially the yttrium aluminum borates. This work may contribute to some good understanding of the interrelationship between the fracture paths and the ballistic behavior. One area that has not yet been addressed is the importance of incoming powder qualities (particle size distribution, particle shape, purity, and vendor variability) of the boron carbide itself as well as the additives.

More attention should be placed on these powder characteristics and their effects, perhaps through using a round-robin or other comparison matrix techniques. Modeling is vital for supporting these experimental approaches to microstructure/properties behavior. While glassy grain boundary phases have been effective in some ceramic materials systems (for example, alumina), they have not been effective in others (for example, titanium diboride), and detailed modeling to understand the relationship between microstructure and observed failure mechanisms is vital information needed for improving structure/property relationships in the future. The development of phase diagrams for any new grain boundary materials will be important, including careful examination of the effects of unreacted carbon and oxygen content on reaction rates and products.

Polymers

Polymeric materials are an important investment area, and their support clearly needs to be continued. In general, this program is doing very well, and it has been recently invigorated through the recent starts of two MCoEs. Perhaps a Multidisciplinary University Research Initiative (MURI) Broad Agency Announcement in this area would be warranted in order to spread the investments across national leaders in the field.

Many of the WMRD briefings were delivered by postdoctoral researchers. This is a plus in that it shows the success of the personnel development strategy being pursued by this program and by WMRD in general. However, it could be a potential negative if it is indicative that the majority of the 6.1 research is being performed solely by the junior researchers. It is understandable that more senior personnel are involved with 6.2 to 6.4 category research and the transitions to the engineering scale-up of the new technology. Senior personnel need to retain a role in the 6.1 research so that they maintain sufficient knowledge of the cutting edge as well as contribute as mentors to the more junior staff and postdoctoral researchers.

The polymer area appears somewhat unique in that it has a strong computational component to support and guide the research. However, the computational polymer science work appears to have been used only to confirm the experimental results. It is suggested that an extension of this effort is needed if the end goal of developing physically based predictive capability is to be achieved. In particular, there needs to be a clear infrastructure for maintaining models, both with respect to their parameterization and their validated regimes.

The projects involved with the design of polymeric materials and nano-engineered additives in order to control morphology and segregation are also a strong component of the research effort and clearly of high merit.

The two Materials Centers of Excellence (MCoE) involved in the polymers area are discussed below.

Materials Center of Excellence—Virginia Tech Because the Virginia Tech MCoE program has been operational less than 1 year, it was difficult to assess. It appears to have a good commitment to outreach and collaboration and is focused on Army needs. This MCoE supports an extensive list of research programs, which in some cases are quite mature. However, the presented research appeared to have been performed using leveraged support provided by other agencies. The research actually supported by ARL's MCoE needs to be articulated, and clear milestones should be established. Without this, it is at present nearly impossible to determine the impact of MCoE funding on the research at Virginia Tech. The topic of impact resistance of polymers was a particularly potent example of the critical need of integrating a strong modeling component into the program if chemistry, formulation, and mechanical behavior are to be linked in a predictive sense. With regard to the program's computational effort, it appears that several investigators are active in this regard, but this was not elucidated in the presentation.

Materials Center of Excellence—Drexel University Because this program has also been operational less than 1 year, it too was difficult to assess. However a few salient points were evident. The program contains promising proposed linkages of synthesis, design, and multiscale modeling and simulation. The list of faculty appears to be dominated by the primary principal investigator (PI) working in conjunction with a few younger PIs. While the PIs are clearly skilled, it was unclear how their various research topics/targets work together with each other. The effort in the computational component needs to be strengthened because it currently contains only one PI involved with only one of the nine research topics. The use of subcontracts to add more expertise could be considered. The center shows good promise as a means of promoting the interaction of graduate students and postdoctoral researchers with ARL. The annual funding level of $500,000 covers nine distinct research topics. This level of support is far too low to expect serious research accomplishments across the several topics. The program needs to focus on a small number of scientific topics encompassing high leverage to WMRD programs and Army needs.

Computational Polymer Science The program in computational polymer science is developing a multiscaled modeling effort that makes use of available software in a standard fashion. The efforts here are the only evidence of serious computational work within this program. The work involves a number of models that span a large range of length and time scales. It is troubling that only one person appears to be working in this effort, because it will likely take a larger commitment to succeed.

Some of the work is first rate—for example, the use of density functional theory (DFT) to design chromophores. This program profits from the close interaction between synthetic and computational chemists and from the comparative ease with which the theoretical results can be interrogated experimentally. Much of the modeling at mesoscopic length scales beyond the quantum level is assumed—but not verified—to be correct through the calculation of various macroscale physical properties, for example density. Rigorous experimental interrogation of models should be built into the program as milestones.

The integration of these models in the coarse-graining direction seemed ad hoc, and a coarse-graining strategy was not described. This leaves open the possibility that successes are accidental and that one might draw poor conclusions when tackling new areas. The information flow in inverting the coarse-grained projection from large scales to small scales was not addressed or even acknowledged as necessary. This is not entirely surprising, because it is difficult to do well, and few people work in that area. The University of North Carolina may be a valuable resource to tap.

That said, the computational polymer science effort is important and should be encouraged in a number of ways. First, there is need to support optimizing these models on available high-performance architectures. Second, this effort appeared to be something of a one-man show. The success of this enterprise hinges on its ability to answer technical problems in a timely fashion (i.e., fast relative to experiment). To do this well and often, simultaneous efforts at parameter tuning, model verification, integration, and system validation are required. To do this on an ongoing basis requires more than one person or a small group. This may call for an unusually large commitment, from the ARL perspective, yet it is necessary for success.

Recent Advances in Selectively Permeable Membranes The work in selectively permeable membranes is a good technologically driven effort that could be aided by science-based rationale. For example, one needs to uncover the chemical mechanism through which barium ions operate to improve selectivity. Is it simply a size effect, or are there more subtle aspects to the phenomenon? The science base can, in part, be supported by computation, but only if augmented by carefully defined experiments. This process will prove to be time-intensive but should result in a long-term payoff.

Nano-Engineered Additives with Self-Stratifying Characteristics Self-stratification of reactive materials on the surface of polymer films supports the desire to achieve spontaneous decontamination of the surface. Two rather clever approaches were presented to control segregation in a composite matrix. The first is a method in which ligands are attached to gold and silicon dioxide particles. The free end of the ligand is designed to undergo a Diels-Alder reaction. This thermally reversible reaction is used to tune the properties of the matrix and the location of the particles by migration. The increased concentration of particles at the surface affords the opportunity to enhance the reactivity, control defects, and change the morphology of the film. The second approach is the use of functionalized hyper-branched polymers to segregate at the surface. By functionalizing the end groups with quat salts, biguanides, alkanolamines, or N-halamines, selective degradation of biological and chemical agents can be achieved. Both of these projects are creative ideas with some potential applications both within and outside the Army.

Analysis of Adhesively Bonded Ceramics The study analyzing adhesively bonded ceramics successfully demonstrated that the toughness of alumina/epoxy bonds is enhanced by grit-blasting the alumina prior to bonding. Additionally, the strength of the alumina substrates is not appreciably degraded, despite the surface roughening and the potential for generation of strength-limiting flaws. These results should provide useful guidance for fabricators of armor systems. The presentation clearly suggested that the principal motivation of this research was to provide industrial guidance in support of Future Combat Systems (FCS) and to provide the basis for techniques for quality assurance.

The toughness enhancement appears to be the result of blasting-induced surface roughness and the ensuing mechanical interlocking of the epoxy with the alumina during bonding. The effects may be anomalously high in the test configuration used to make this assessment. That is, in the asymmetric wedge test, the crack tip stress field contains a significant mode II component. This component is likely to be largely shielded from the crack tip when crack tortuosity is high. Efforts to measure the mode I toughness may prove useful, since this mode invariably yields the lowest value (relative to those for mixed mode I/II cracks) and is less strongly influenced by roughness. It is understood that finite element modeling (FEM) of the asymmetric wedge test is currently underway in the Survivability program, and WMRD should continue such an effort to facilitate future predictive capability in support of complex composite armor design and testing.

Surface Modification of Ultra High Molecular Weight Polyethylene (UHMWPE) This program is examining how the strength and durability of glass-reinforced composites can be improved through manipulation of the texturing of the fiber surface. Enhancement of the strength and energy absorption of the fiber-matrix interface is a critical research area for both glass and polymeric fibers. The atmospheric plasma treatment of both UHMWPE films and fibers clearly suggested potential for improvement, based on oxidation of the surface and its beneficial effects on fiber bonding. Similarly, the plasma treatment and its effects on wettability when silica is deposited on UHMWPE fibers clearly indicate improved adhesion and improved mechanical properties. There are clear signs of scientifically inspired approaches to improving the engineering properties of UHMWPE in composite applications, but WMRD should examine previous R&D on strength/wear/friction research in the biomaterials literature following the use of plasma modification of UHMWPE in implant applications and the use of surface modification.

Lethality

Affordable Precision Munitions

The Affordable Precision Munitions program is an outstanding achievement for WMRD. In this program a multidisciplinary design approach is being applied to the development of small, compact, guided munitions. The multidisciplinary approach employs such novel features as the following: (1) virtual wind tunnel tests and virtual fly-outs that allow a much larger range of designs to be explored at a much lower cost than would be the case for real wind tunnel tests; both significantly and surprisingly, results have been achieved in terms of nose and body interactions, jet interactions, and diverter interactions that have permitted avoidance of unstable designs; (2) clever use of piezoelectric actuators that provide guidance control; and (3) even though the Global Positioning System would seem to be the first choice for guidance, some out-of-the-box thinking suggested that in this case space constraints and battery-drain considerations would be better accommodated using magnetometers to provide guidance information.

Scalable Technology for Adaptive Response

A new-start research effort that shows considerable promise is that of Scalable Technology for Adaptive Response (STAR), which examines the development of weapons that can be tailored to deliver a spectrum of weapons effects, including different yields, controlled fragmentation, selectable fragmentation, and behind-armor and threat effects. The program objective of pursuing the development of single munitions capable of addressing multiple mission capabilities rather than requiring a spectrum of munitions in-theater is an innovative and rational goal. The STAR program entails a synthesis of expertise in shock physics, manufacturing technology development, complex fusing, and selectable material response (such as is possible with shape memory materials). This project embodies a system approach to a warhead development that is to be encouraged within WMRD.

Missile Propulsion Modeling

WMRD's missile propulsion modeling efforts are aimed at understanding selectable trajectory control in liquid propellant thrusters. These thrusters apply to control over in-flight projectiles. Hypergolic fuel and oxidizer mix in the combustor under conditions that allow the mix to be pumped as needed by the stage of flight. The work employs a reactive computational fluid dynamics (CFD) model that was developed primarily at ARL. The research group conducting this work has had extensive experience in modeling solid gun propellant combustion and is extending this experience to liquid propellants. On the downside, the codes seem to be very inefficient and are usable because of the massive amount of computer power available. The research group has broad collaborations with industry, universities, and other laboratories and is receiving a positive reaction from the user community. If not already done, it would be wise to review the research that the Air Force Office of Scientific Research supported in the 1980-1995 time frame in the area of instability in liquid propulsion systems.

Multifunctional Warheads

The program on multifunctional warheads, in its last year as currently configured, is developing a multifunctional munition in which kinetic energy (KE) and behind-armor effects and/or blast are considered. The focus is on how to harness rocket propellant to increase the engagement velocity of a munition after ballistic launch against different threats by employing a detonable propellant, specifically a hard target (concrete walls or three-layer brick), and/or to allow munition-threat interaction and thereafter to use the rocket propellant to achieve increased blast and/or increased behind-armor effects. This technical approach, like ARL's new initiatives in tunable/scalable munitions, represents a future-looking, technically challenging approach to novel munition development and is to be encouraged. A continuation of this project's technical approach in the future by way of transition to other applications (smaller-caliber threats) and/or through a follow-on project continuing this research direction appears technically warranted and is encouraged.

Military Operations on Urbanized Terrain Lethality

Munitions directed at typical buildings in urban environments often penetrate walls and produce secondary fragments. These fragments can cause significant collateral damage, including the injury or death of noncombatants. The Military Operations on Urbanized Terrain Lethality program is designed to collect data from the impact of standard munitions on walls typically found in urban contexts (poured, reinforced concrete; concrete block; brick; and other materials). These data are used to inform the development of computational models for predicting the amount and distribution of fragmentation produced by selected munitions interacting with selected building materials. These models can be used to evaluate munition design as well as protective measures for the warfighter who may use urban structures for shelter.

This is valuable work that addresses real problems in current and future theaters. The project will also use the results of the soft-tissue modeling effort at WMRD. Collaborations (e.g., with Germany) are ongoing. The project requires a significant amount of labor to collect data; this effort is primarily done by one person. This project is significantly understaffed, especially given the labor-intensive nature of debris field data acquisition. Expanded staff would allow a wider range of munitions and building materials to be investigated. Increased interaction with the soft-tissue modeling effort would be useful to that project as well as to related efforts underway external to WMRD.

Sensor, Warhead, and Fuze Technology Integrated for Combined Effects

The project Sensor, Warhead, and Fuze Technology Integrated for Combined Effects is a portion of the WMRD's larger multifunctional warhead and munitions effort, focused on the ability to put sensors on the end and forward surface of a munition to sense the type of target and then to enable selective fusing in order to tailor the munition to that target. The goals of the larger project are to simplify logistics by stocking only a single munition type rather than having to estimate the likely target types prior to loading, to automate target discrimination and fusing, to reduce human error, and to enable more rapid fire by the soldier.

The technical challenges include determining the most appropriate sensor types, determining the materials for the end cap and the sensors that can survive initial impact, and developing a selective fuzing methodology. The team performing this work is also assessing possible spoofing threats and whether it is possible to readjust the fuzing in flight as more data are received on the sensors. This work will combine with the scaled effects endeavors in the larger project. The overall project will also be evaluated from a systems perspective to determine that the solution is worth implementing. This is an important and intriguing project, and while it is in its initial stages, it appears to have an appropriate overall project plan.

Electromagnetic Gun

This project for an electromagnetic gun showed great advances in materials manufacturing and complex engineering. Challenges of heat buildup are being addressed with active cooling, and firing demonstrations have been successful. The largest challenge for the Army with this gun is the power supply for a mobile gun. The ARL work needs to be closely coupled to the development of the counter-rotating motor power generation program at the Army's Armament Research, Development and Engineering Center (ARDEC). This is a high-risk project requiring success in projects at both facilities. It might be appropriate to consider a risk-mitigation strategy if the motor proves insufficient for the task. Close coupling with the Navy's program is also appropriate, and the investigators seem to be aware of what is going on broadly throughout the nation. The intriguing idea of a hybrid EM and conventional gun (propellant-driven) is also being pursued, and results of the ongoing paper study should prove interesting.

DEMN (Explosive Fill)

The project on DEMN (an explosive fill) is an example of excellent directed engineering that builds on unique capabilities recently brought online at ARL such as the Novel Energetic Research Facility. The development of an insensitive munitions explosive material, with performance, manufacturability, and cost near that of TNT, is a well-defined goal. The development of DEMN also demonstrates a continued paradigm shift from looking for one magic explosive material to using calculations and experience to suggest a mixture of materials that will provide the needed performance. The demonstration of mild response to sympathetic detonation without barriers was a major accomplishment. The team at ARL is well situated to advance this science, having computational expertise, explosive formulation facilities, and testing capabilities. Also, the investment in the NERF facility is commendable because it places ARL in a unique position to fill a need for formulation and scale-up testing. Challenges remain in monitoring the performance of materials when scaled up at commercial sites, because manufacturing processes can affect explosive sensitivity. This program demonstrates a mix of theory, computation, and experimentation. The largest scientific challenge in the continuation of this type of work is the development of a predictive capability for the transient behavior (ignition and failure) of non-ideal explosives. The fidelity with which the chemistry and physics of reactive wave growth are described is higher for predicting failure diameters than is needed for performance parameters such as detonation pressure and velocity.

Related to the property testing of the DEMN insensitive munition is the effort on modeling munition response. The computational work focused on the use of the CTH code to model shock-initiated failure.[1] While CTH provided good mechanistic understanding, it was not quantitatively valuable, and the ultimate performance numbers were (appropriately) derived from experimental results. It is clear that very long time failure processes, like slow cook-off, are not yet amenable to predictive modeling. However, the expectations for shock-initiated processes are greater, and the inability of CTH to validate experimental results is of some concern. What seems to be missing is an active feedback loop whereby these computations would inform improvements to the code or constitutive models. Absent this feedback, it is not obvious that the existing modeling capabilities can play a meaningful role in munitions design and characterization.

[1] The CTH code, developed by the Sandia National Laboratories, provides capabilities for modeling the dynamics of multidimensional systems with multiple materials, large deformations, and strong shock waves.

Theoretical Chemistry and Advanced Energetic Materials

The goals of the theoretical chemistry and advanced energetic materials effort are projected to be as follows: (1) the development of high-performance computing capabilities necessary to characterize energetic materials, including the identification of fundamental mechanisms for the control of reactions; (2) improved thermodynamic predictions; and (3) the use of computation as a means to discover new ways to store and release energy. Currently, some effort is being made to establish underlying mechanisms controlling shock-initiated chemistry, by targeting molecular energetic crystals and, specifically, effects of pressure on properties, and controlling the associated intermolecular forces that are responsible for the properties being investigated. The use of modeling for the prediction of heats of formation is fairly well established, and while the group performing this work does not give the impression that it works toward development of its own methods, it is clearly consulting or collaborating with top-notch groups that do. In particular, this is enabling the group to benefit greatly from the latest state of the art in density functional techniques, such as those offered by the newest dispersion-enabled functionals and pseudopotentials of Grimme, Truhlar, and Rothlisberger. On the other hand, a search into the literature indicates that there is work going on (also coupled to some long-standing collaborations of experts in the community) in the development of capabilities specifically to characterize energetic materials, emphasizing the prediction of properties associated with the performance and sensitivity of materials. This has involved such key developments as the following: the SRT model, which is constantly being refined and extended, in particular now with the advancements in the density functional theory models; and methods for predicting properties of energetic molecular crystals, which has long been a difficult area but for which there are new methods such as those being explored in this group and, for example, a similar model, PIXEL. However, there are clearly some advantages that the group could gain by fostering its own code-development skills directly as a part of the group's activities.

It appears that the theoretical chemistry and advanced energetic materials group does not take advantage of other computational and theoretical development skills within other groups in ARL. Several discussions suggested the need for improved bridges from the molecular scale. The theoretical chemistry group needs to be given a stronger role in these efforts. At present, it appears that the other efforts are developing their own theoretical chemistry expertise without much engagement of this nature. As the theoretical chemistry group develops such ties, it would presumably build some 6.2 and 6.3 components. Certainly, there is a lot to be gained by such collaborative activity in terms of improving the group's code-developing skills and having the advantage of being directly tied to the experimental direction, something that appears also to be a weak component of the theoretical group. In particular, there is a need to develop multiscale tools that connect the codes that are being used across ARL, from the molecular to macromolecular scales of the theoretical chemistry group to the materials and engineering scales in the other groups. For example, multiscale molecular dynamics modeling and crystal/molecular packing, using the high-level data offered by quantum mechanics, should be considered for the purpose of analyzing the shock compression and shearing sensitivity of materials under extremes of pressure and temperature. In several cases, gaps in certain parameters at the higher scales described during the presentations could be filled in using quantum mechanics data.

Materials for Lethality

The current focus areas within WMRD's program on materials for lethality are tungsten for KE munitions, structural reactive materials, and cold-spray particle disposition. The program centers on novel and/or discovery science and concentrates on addressing more near-term applied engineering problems.

Replacement of DU munitions with W-based materials remains the central goal of this effort. The pursuit of pure W-based materials manufactured using powder metallurgy or equal channel angular pressing (ECAP), each singularly or followed by postprocessing (rolling), appear to offer promise. The question of scale-up appears to be worthy of consideration in both instances—powder metallurgy or ECAP. A combination of integrated (ballistics) and fundamental characterization (high-rate constitutive and fracture toughness testing) is encouraged as soon as possible during the assessment of these materials.

The structural reactive materials (RMs) effort presents a compelling focus on developing and manipulating microstructural aspects of RMs to simultaneously achieve multifunctionality (structural capability and reactivity). The coupling of modeling with experimental efforts is encouraged in addressing these coupled goals. The fabrication of unique and tailored materials not obtainable by other processing routes using cold-spray manufacturing was described. Exploration of this technology to produce near-net shape components, such as rocket nozzles or munitions, offers some intriguing possibilities.

Further, the use of cold spray as a fabrication route to the production of powders to support the novel W-based materials or reactive materials projects appears technically productive and a novel route to complex alloy powder production. In the area of cold spray, progress has largely been in new materials (tantalum has been very successful, but molybdenum has oxidation issues; the real goals are tungsten or copper). There has also been progress in terms of heat treatment of the deposited layers and testing to demonstrate wear resistance and adhesion of the cold-spray layers. The researchers did not describe any fundamental understanding of why cold spray works or of what the local microstructure or physics effects might be. It is possible that this fundamental work is underway and simply was not described; if it is not underway, it should be.

Materials Under Extreme Pressures

The work on materials under extreme pressures is focused on exotic energetic materials, specifically equations of state (EOSs) and phase diagrams using the diamond anvil cell. There is an important interaction with the Carnegie Institution of Washington, where outstanding and successful research at high pressures has been conducted for a long time. The impressive EOS and phase diagram studies are important and useful contributions to fundamental aspects of energetic materials. Caution should be exercised with respect to how the work on exotic energetic materials such as polynitrogen and processes such as structural bond energy release is sold. These latter subjects are provocative and push the extreme of chemical imagination. On the one hand, they risk tying the investigator to the growing list of painful attempts to challenge the laws of nature. On the other hand, proven success in one of these areas assures instant fame, however impractical the result might be. It is important to be mindful that the research is still limited by thermodynamics laws.

Reactive Material Energy Release Mechanisms

The work on RM energy release mechanisms is focused on the development of modeling tools and experimental chemical measurements directed at understanding the energy release by reactive materials in conjunction with explosives. The approach is to understand how to tailor energy release, to model the macroscale, and to develop appropriate diagnostics. More specifically, modeling of the shock wave/fireball, fragment formation, and the effect of additives is being conducted. Experimentally, laser-controlled initiation, the emission spectroscopy of product species, and unique spatial and temporal thermometry are focal points. Blast enhancement by surrounding the explosive with RMs such as Al and Ni alloys has been observed and is being investigated. Important Army applications include the development of thermobaric explosives and RM shell cases. In general, the work fits Army needs and is well done.

Reactive Materials

Reactive materials can deliver chemical energy far in excess of that available from impact alone. Research in this area is essential to ARL's mission and will result in greater lethality. Fundamental investigations of RMs will involve the quantum chemistry group, and although there is research going on in this area (quantum chemistry and reactive materials) it does not, on the surface, seem to be influencing the development of RMs. The project encompasses both engineering approaches to deliver near-term product (building on previous work in the Navy and elsewhere) and longer-term research to predict and evaluate the performance of other materials (including original work at ARL). The RMs are not explosive but do, on impact, deliver some degree of energetic output relative to an inert projectile. Many of the challenges relate to the mechanical behavior of the RMs: strength needed for launch and flight, density, and manufacturability. Ongoing research to pursue materials with improved mechanical performance with enhanced reactivity is well directed and seems to include appropriate simulation and experimentation.

Survivability

Reactive Armor

Reactive armor (RA) is a common form of add-on armor, used on many armored fighting vehicles. This is a proven concept first used by the Israel Defense Forces successfully in combat with the Israeli Army M-60s and Centurion tanks in the 1982 War, and later by the Soviet Army in the mid-1980s. The RA concept employs add-on protection modules consisting of thin metal plates and a sloped explosive sheath, which explode when sensing an impact of an explosive charge. The RA enables a significant increase in the level of protection, primarily against conventionally shaped charges and the explosively formed penetrator, a special type of shaped charge designed to penetrate armor effectively at standoff distances.

The ARL use of modeling and simulation of RA to help understand the impact of several key variables is excellent. The researchers have made outstanding strides to understand the physics integrated with the ballistics to make the current RA successful.

The arrival of ARL-recommended RA kits in-theater will be the ultimate verification and validation of this program and of the current modeling and simulation methodology. ARL needs to ensure that the data and experiences of the soldiers using the RA package are given to the developers, closing the loop.

Transparent Armor

WMRD's future-force program is focused on the evaluation of transparent materials and on WMRD's approach to design solutions to meet the needs for transparent armor, sensor protection, and glazing life protection. The project on transparent armor is specifically focused on the following: materials and laminate design, ballistic design, and ballistic modeling. The current materials under consideration are mostly in-hand materials, and the focus is on increasing protection without an increase in weight. Realization of this objective while balancing the requirements of optics, transmission, ballistics, scratch resistance, and solar loading makes this area a challenging one in which ARL clearly has established expertise and vision for the future. Details of the solar loading and challenges in thermal management toward the development of a ballistics specification are experimentally fascinating and show a healthy balance of experimental and modeling focus. The data clearly support conclusions of the importance of needing to quantify transparent armor performance above ambient temperatures. This program evinces experimental and modeling balance coupled with near-term application-driven tasking and a long-term vision of where transparent armor evolution needs to proceed to support future force needs.

Ceramic Armor Materials

The work on ceramic armor materials is a piece of the larger work on ceramic composite armor. The stated goal is to achieve a fundamental understanding of deformation and failure mechanisms at ballistic conditions. There is a lack of good data and models at these very high shear rates, and it is an important problem to work on. As presented, the experimental work to date seems very empirical and exploratory. It is unclear whether existing literature has been leveraged sufficiently. While it is probably true that data do not exist at ballistic conditions, there is ample literature at slower rates, which could inform the choices of experimental techniques and limits.

Composite Ceramic Armor Performance

The underlying failure mechanisms occurring in composite ceramic armor components were identified through postmortem analysis of representative impacts, and strategies to rectify these problems through modification of the binder and the rigid components were described. The binder study appeared to be systematic, with the key properties of candidate binders being characterized and well understood. Some innovative changes were proposed for the structural components, and these innovations appear to be very successful. It is not clear that a systematic plan exists for refining these new strategies.

Computational approaches were suggested by WMRD, but predictive computational capabilities for such complex, textured composite materials were not demonstrated. Some simulation of fabric modeling alone was presented, but the relevance of this modeling seems directed to manufacturing and appears to be disconnected from the overall composite behavior in impact.

Electromagnetic Armor Physics

The program on electromagnetic armor physics emphasizes the mechanism by which EM armor addresses jets and the role that computational validation plays in this process. The contrast to challenges presented by explosively formed penetrator devices was described. The Alegra code enhancement and concurrent physics hypothesis testing that resulted from the jet validation work (specifically, air conductivity) are commendable. However, a similar attention to fracture was absent and represents a significant lost opportunity. The need for predictive physics-based fracture models was a recurring theme in the overall WMRD presentations.

Tactical Wheeled Vehicle Survivability

As in the case of the RA program, which is addressing current Army needs, the Tactical Wheeled Vehicle Survivability program is an excellent example of the duality of ARL's missions to simultaneously quickly solve in-theater problems to support the warfighter while considering how to effect changes in armor materials on future systems. ARL's innovative and rapid turnaround approach to up-armoring door panels for tactical wheeled vehicles in response to needs from the warfighter is commendable. Not being satisfied with resting on its laurels, ARL is continuing research on new materials for lightweight tactical wheeled vehicle (TWV) armor applications in the future by looking, for example, at the ALCAN aluminum alloy 2139. Both the motivation and the research approach effort are to be applauded as clearly technically driven and forward thinking, with the goal of achieving improved performance in future TWVs. ARL is encouraged to work with program managers and prime contractors to orchestrate the insertion of the new aluminum alloys showing promise into future platforms and as replacement materials for upgrades of existing TWVs.

Kinetic Energy Active Protection Systems

The program on kinetic energy active protection systems is intended to deploy an explosive projectile that will detonate near incoming KE penetrators and cause them to swerve, missing the intended target. The multimedia demonstrations were excellent and provided a realistic sense of the complexities involved in the design and deployment of an active protection system for KE penetrators. This is primarily an engineering program utilizing sensors to identify the friction-produced heat signature of the incoming KE projectile and then maneuvering to the penetrator. The part of the program involving the identification of the incoming KE round is near completion. There is every reason to expect that an active protection system will be deployable in the near future.

Composite Materials Technology for Armor

A very detailed model of the fibers, threads, and weave of the fabric for future composite armor components was presented. These models will be very helpful to the future manufacturability of the fabric. Simulations of impact damage were also shown, but the overall project plan (including the requirements and selection of adhesives and matrix materials) was not clearly articulated.

Armor Technologies for the Current Force

The effort involving armor technologies for the current force is primarily driven by expedients, and the WMRD team appears to be satisfying that need well as well as integrating armor technologies with existing and emerging systems. The group has actively pursued the use of many new materials and engineering designs to improve the efficacy of armor technologies. This is also visible in the products that it has delivered to the field. Although the speed with which this group is exploring new materials makes it very hard to implement them into simulation codes, the group is attempting to do so. This is thus a good case study illustrating the need for the 6.1 materials computing area referred to above to develop transferable constitutive models capable of describing new materials quickly.

Pulsed-Power Armor Technologies

Future Army technologies will require increased flexibility, function, and extension, as well as support for unconventional weapons and armor systems, such as demonstrated in pulsed-power armor technology. Modular pulsed-power technologies have the potential to be used in many applications and field implementations for defense, making this area crucial to ARL. To meet these requirements, this demonstration showed advancements to support propulsion, continuous auxiliary power, and pulsed-power demand for weapons and armor. One question would be whether the power supply needed for these purposes can be supported for general use.

Predictive Capabilities for Buried Blast Threats

The effort involving predictive capabilities for buried blast threats is focused on experimental results for buried threats, with particular emphasis on the role of burial depth and soil properties. Scaling laws were proposed, and some simulation results were presented that support the experiments. These simulations incorporate new constitutive models developed through academic collaborations. Quantitative modeling of soils is very difficult and seems appropriately targeted as a research avenue.

Vertical Impulse Measurement Facility

The Vertical Impulse Measurement Facility provided data necessary to tune and validate models simulating the impulsive output of buried charges and the response of targets of interest, particularly the vertical impulse from buried charges weighing up to 8 kg. This is a program that reflects the general ARL philosophy of validating models to the level necessary to provide designers with confidence in the results. This facility is crucial to the ARL mission and has been used productively in support of that mission.

Novel Energetic Research Facility

The Army has made a wise investment in rebuilding the Novel Energetic Research Facility for the pilot-scale formulation of explosives. The amounts of explosive that can be obtained enable engineering-scale tests to be conducted. The hiring of additional employees to work in this facility is laudable and helps fill a national need. The success of DEMN as an insensitive explosive fill is a useful advance to justify the investment and provides a selling point for additional growth in this area. The scale-up and formulation of explosives are usually the "valley-of-death" for many explosive programs. There has been an overall disinvestment throughout the nation in such facilities, which places ARL in a unique position to fill a critical need for the energetics community.

OPPORTUNITIES AND CHALLENGES

Predicting High Strain-Rate Properties

WMRD presenters noted that in the world of armor protection, scientists and engineers speak of horrendously large forces being applied to materials in microseconds or faster, a phenomenon known as high strain-rate deformation. At these high strain rates, materials properties can be markedly different from what they are at slower rates, such as, for example, when tested in an Instron. Accordingly, when developing new armor concepts it makes little sense to use handbook values of materials properties such as tensile strength, yield stress, or fracture toughness, because these could be orders-of-magnitude different from high strain-rate properties applicable to armor. Fortunately, however, there is a way of obtaining such high strain-rate properties using Hopkinson bar experiments in which flyer plates are accelerated toward small samples using high-pressure gas jets. Unfortunately, the experiments tell nothing about what metallurgical factors control properties under rapid deformation, so there does not exist today any way to design materials having preselected high strain-rate properties. The high strain-rate world therefore differs markedly from the more familiar low strain-rate world in which materials scientists often know what is needed to increase the yield stress or fracture toughness of many advanced materials that are to be used in more normal applications. In the world of high strain rates, however, very little, if anything, is known about tailoring the strain-rate sensitivity of potential armor materials, leaving the design engineer in the dark and the ability to design armor for some intended application very difficult.

The development of materials for low strain-rate applications has benefited in the past from many years of intensive efforts by materials scientists, but currently as new and more demanding applications appear on the horizon, some reliance is being placed on computational materials science in which techniques such as molecular dynamics simulation and electronic structure calculations are employed to understand how features at the atomic, mesoscale, and microscale levels influence the structure/property relationships.[2] Seemingly, no such effort has been made to focus the attention of computational materials science (CMS) on the problem of understanding what factors govern the markedly huge difference between the properties of many materials at high strain rates and their better-known properties at low strain rates. Perhaps the reason is that such a study would have to cover many length and time scales and would have to deal with the dynamics of the situation as well as other factors that undoubtedly control structure-property relationships in the high strain-rate regime. In short, it would not be an easy task. However, a program designed to develop such an understanding should be of great benefit to the armor community in that, if successful, it would for the first time allow these researchers to tailor the all-important high strain-rate properties of a potential armor material to specific armor applications. This would constitute both a challenge and an opportunity for ARL.

Computational Materials Science

As noted above, the balance between experiment and computational efforts in materials R&D has improved markedly within WMRD over the past 2 years. Nonetheless, the development of CMS must be undertaken within the context of a clear objective. The challenge for WMRD, therefore, is as follows. Typically, the objectives of institutional CMS efforts fall along a continuum. At one end are those that push the computational and theory envelope; at the other are programs that support or are integral to the materials development process. The goals in the first of these extremes is to develop or accelerate computational tools, while for the latter extreme the goal is to explore how existing CMS tools can be used to hasten the materials design and deployment process. Considering finite element modeling (FEM), there is good work proceeding at both extremes: computational scientists are exploring ways to expand FEM to larger systems, while at the same time scientists and engineers are employing these techniques to design new structures and processes. Example programs drawn from across the spectrum of CMS efforts range from those at the California Institute of Technology, which can be classified as pushing the computational and theory boundary, to those at Northwestern University and QuesTek Innovations LLC, which have integrated CMS into their materials design process. In the middle is the group in the Materials and Manufacturing Directorate of the Air Force Research Laboratory at Wright-Patterson Air Force Base.

The key differences between these extremes are the skills of the personnel involved. Usually, computational scientists and theorists, who for the most part work independently of experimentalists, staff a program pushing the CMS envelope. At best, these researchers are called on to explain experimental

[2] See, National Research Council, National Materials Advisory Board, *The Impact of Supercomputing Capabilities on US Materials Science and Technology,* Washington, D.C.: National Academy Press, 1988; and National Research Council, National Materials Advisory Board, *Integrated Computational Materials Engineering: A Transformational Discipline for Improved Competitiveness and National Security,* Washington, D.C.: The National Academies Press, 2008.

observations. For CMS programs directed at materials development, experimentalists and materials scientists use existing software, much as they now use microscopy or spectroscopic information. Given the resources and other constraints placed on ARL, its mission is best served through the development of a CMS effort patterned more on that of the group at Northwestern University than on the California Institute of Technology group.

An effort patterned along the lines of the Northwestern paradigm is "grown locally." As a materials development program is planned, a redundant modeling and experimental effort is created. As an illustration, if one is exploring the effects of alloying or processing on a well-characterized property, a set of experiments is performed in which the processing parameters or alloying elements are varied and the property is measured. For such cases, a redundant CMS effort should operate in parallel with the experiments. In this way, experimental verification of the models becomes an integral part of the design effort. At the same time, the experimentalist learns how to incorporate CMS results into his or her materials development programs. Over time, the experimentalist will gain confidence in the models just as one gains confidence with experimentally derived information. The work at ARL on chromophore design is an excellent example of this type of research. The modeling and experiment are well integrated, and both are targeted to single property prediction.

Soft-Tissue Physics

The new program on soft-tissue physics, whose motivation is provided by the need to better understand the interactions of munitions, directly or indirectly, with humans, represents an opportunity for WMRD, since it is in an area of great interest in Congress and among the general public. It also represents a great challenge, because it is unlike any prior programs in WMRD and will therefore have to undergo a somewhat extended learning curve. The primary concerns of the soft-tissue physics program are penetrating wounds from munitions (these may be from projectiles or fragments), blunt trauma, and blast loading (shock wave effects, especially, traumatic brain injury).

The current focus is on the development of computational models of soft-tissue response to high strain rates induced by projectiles and shock waves. For empirical input into model development and validation, the program is examining historic data from the United States Army Wound Ballistics Laboratory at Edgewood Arsenal, Aberdeen Proving Ground, Maryland. In addition, WMRD is conducting high strain-rate experiments with tissues using the high-rate split Hopkinson pressure bar and is attempting to design clamping mechanisms to allow strain-to-failure experiments with tissues. Shock physics codes from the Department of Energy are being used (and adapted) to support the modeling of blast loading. Model development is directed at the torso (customer-driven work) and the brain.

WMRD's successful efforts in refocusing this program following the Board's 2006 review are commendable. There is a national defense priority addressed by this program. The understanding of munitions-induced trauma is of high importance in the design of armor, in the design of armaments (e.g., to reduce the risk of collateral damage to noncombatants), in the design and delivery of effective treatment of the wounded, and in the training of medical personnel. WMRD's program is a vital component in this national effort. TBI has a very high profile with the public and Congress; hence WMRD's efforts in modeling blast loading are timely. The problem is large and complex, and many players (from other government agencies, universities, and international partners) are participating in as-yet largely uncoordinated efforts to achieve useful solutions. The historic data from the United States Army Wound Ballistics Laboratory at Edgewood Arsenal are a unique resource that could yield very important (and currently unobtainable) data on trauma. WMRD is uniquely qualified to deal with shock physics and to partner in gathering vital empirical in vitro data for high strain rates in tissues.

There are concerns: The selection of tissue types to be modeled requires consideration (i.e., in the case of a limb, should the model be limited to bone and muscle, excluding nerve and vascular tissue?). The degree to which individual tissue types can be modeled and then the individual models aggregated to form a system model also needs further consideration (e.g., addressing to what extent one type of brain tissue can be modeled and that model successfully integrated with a complete model of the head that encloses the tissue in a cranium and incorporates vascular tissue and other types of brain tissue). The development of anatomical models using grid geometries may not be optimal for use in all application areas and/or easily integrated with other geometrical models. There is an apparent lack of use of decades of research in attaching tissues to fixtures for mechanical tests in favor of developing these attachment mechanisms anew. There is a lack of WMRD personnel with bioengineering/biomechanical backgrounds (although some WMRD external collaborators do possess such backgrounds). There is need to consult with Human Research and Engineering Directorate biomechanics personnel to improve the modeling being proposed and to work with other biomechanics groups outside the Army.

Better cognizance of previous and existing efforts is essential. For example, among projects funded by a Defense Advanced Research Projects Agency (DARPA) program in Combat Casualty Care in the 1990s, one was performed by MusculoGraphics, Inc., which developed a Limb Trauma Simulator incorporating a model of a trauma wound to the thigh.[3] Another DARPA effort is, among other activities, developing a high-fidelity heart model. Another example of extensive work in computational models of soft tissues is that of Montgomery and co-authors.[4] In addition, WMRD should investigate the biomechanics program at the University of California, San Diego. In the same vein, a deeper cooperation with the brain model program of the Massachusetts Institute of Technology would ensure that WMRD model development is complementary rather than duplicative of that effort.

There is need for better coordination of WMRD efforts with those of others. For example, WMRD should consider participation on the Defense Science Board Medical IED (Improvised Explosive Device) Panel. WMRD should develop a systematic approach to data mining the Edgewood data. These data are worthy of a fresh look in light of what is needed and may support validation of development models.

WMRD should consider acquisition of some in-house expertise in the form of at least one bioengineering or biomechanics expert. Such an expert, in cooperation with external medical experts, could better guide the selection of tissues to be modeled, the choice of grid geometries, and the validation of those models with existing or future empirical data.

Models that simply reflect fixed (i.e., one person's) anatomy will not be as useful as those that can represent a reasonable range of anatomical variations such as are found in the general population. Developing anatomical models with this capability should be a goal. The review of empirical anatomical wound data and the compilation of tissue mechanical property data should advance prior to the large-scale development of anatomical models (grid libraries). Until it is known which tissues are likely to be most important, it may be wise to defer anatomical model development. Such a deferment would also provide an opportunity for the investigation of the large number of currently available anatomical models.

Trauma to in vivo soft tissues may do more than alter the geometry of the tissue. In almost all cases, tissue physiology (and that of organs and the organism) will also be affected. Thus, physiological modeling should be included in the overall effort; however, this type of model development may be best done by other organizations and integrated with WMRD models.

---

[3] See, for example, Richard M. Satava, "Surgical Education and Surgical Simulation," *World Journal of Surgery* 25:11 (2001), pp. 1484-1489.

[4] See, for example, K. Montgomery, C. Bruyns, J. Brown, G. Thonier, F. Mazzella, S. Wildermuth, S. Sorkin, A. Tellier, B. Lerman, B. Beedu, and J.C. Latombe, "Spring: A General Framework for Collaborative, Real-time Surgical Simulation," *Medicine Meets Virtual Reality* (MMVR02), Newport Beach, Calif., January 23-26, 2002.

Careful examination of previous efforts to clamp or otherwise attach tissues to fixtures for mechanical measurements should be done.

Model development should proceed with consideration for the likely integration of WMRD models with those developed by other organizations. Such integration requires a careful documentation of model assumptions, prudent choices of grid geometries, and the provision for data interchange with other models.

OVERALL TECHNICAL QUALITY OF THE WORK

The Weapons and Materials Research Directorate conducts activities across a very wide breadth and depth. The fact that even in time of war when heavy demands have been placed on WMRD, the directorate has to respond in the short term to serious problems faced by the warfighter (e.g., up-armoring Humvees) and still has been able to maintain an excellent series of R&D programs to ensure that the warfighter of the future will receive the same benefits. WMRD's slogan, Technology Driven, Warfighter Focused, suggests a top-down organizing principle buttressed by science and technologies derived from internal efforts as well as from various Materials Centers of Excellence at several universities. The interaction of WMRD staff with the MCoEs has benefited the organization by creating a link between the basic research results coming out of academia and the somewhat more applied needs and programs of ARL. MCoEs have also been the source of a number of summer postdoctoral researchers, some of whom have stayed on as staff members. Many of the presenters in the 2008 review were postdoctoral researchers and young staff members who showed the enthusiasm of youth in their presentations. WMRD is encouraged to continue and even to expand these connections to universities.

The experimental work and the computational materials, modeling, and simulation work being carried out are, in almost all of the activities, of high quality. For example, the work employing DFT to develop various nonlinear optical materials including chromophores for use as eye and sensor protection films is an excellent example of good research that profits from the close interaction between synthetic and computational chemists.

The laudable success of the Affordable Precision Munitions program was due in no small part to its emphasis on multidisciplinary design (MDD). This approach combines capabilities from a number of disciplines that in this case included virtual wind tunnel techniques allowing a wide range of designs to be explored, a novel guidance approach, and special propellants. In this case, MDD has resulted in a highly successful program, and WMRD has employed this technique as a means of accelerating development and reducing risk; it should be considered as a model in the future for use in designing other lethality and survivability systems.

The use of virtual wind tunnels and virtual fly-outs allows a much larger range of designs to be explored at much lower cost in much less time. Significant, and surprising, results have been achieved in terms of nose and body interactions, jet interactions, and diverter interactions that have permitted development to avoid costly pursuit of what would be unusable designs. Documentation of model assumptions and ongoing validation of models and virtual wind tunnel and virtual fly-out codes are essential to maintaining confidence in the program results.

Continual refinement of models and their use by various personnel require that all model assumptions be carefully documented. Systematic and selected validation experiments must be conducted in wind tunnels and on instrumented ranges to ensure that models and the performance testing codes are sufficiently trustworthy to support decisions on final designs.

Appendixes

Appendix A

Army Research Laboratory Organization Chart, Resources, and Staffing Profile

This appendix presents an organization chart of the Army Research Laboratory (ARL) in Figure A.1 and data on ARL funding and staffing in Tables A.1 and A.2, respectively. Table A.1 indicates the type and amount of funding by directorate for fiscal year (FY) 2003 through FY 2008, and Table A.2 presents staffing profiles by directorate for the years 2004, 2006, and 2008.

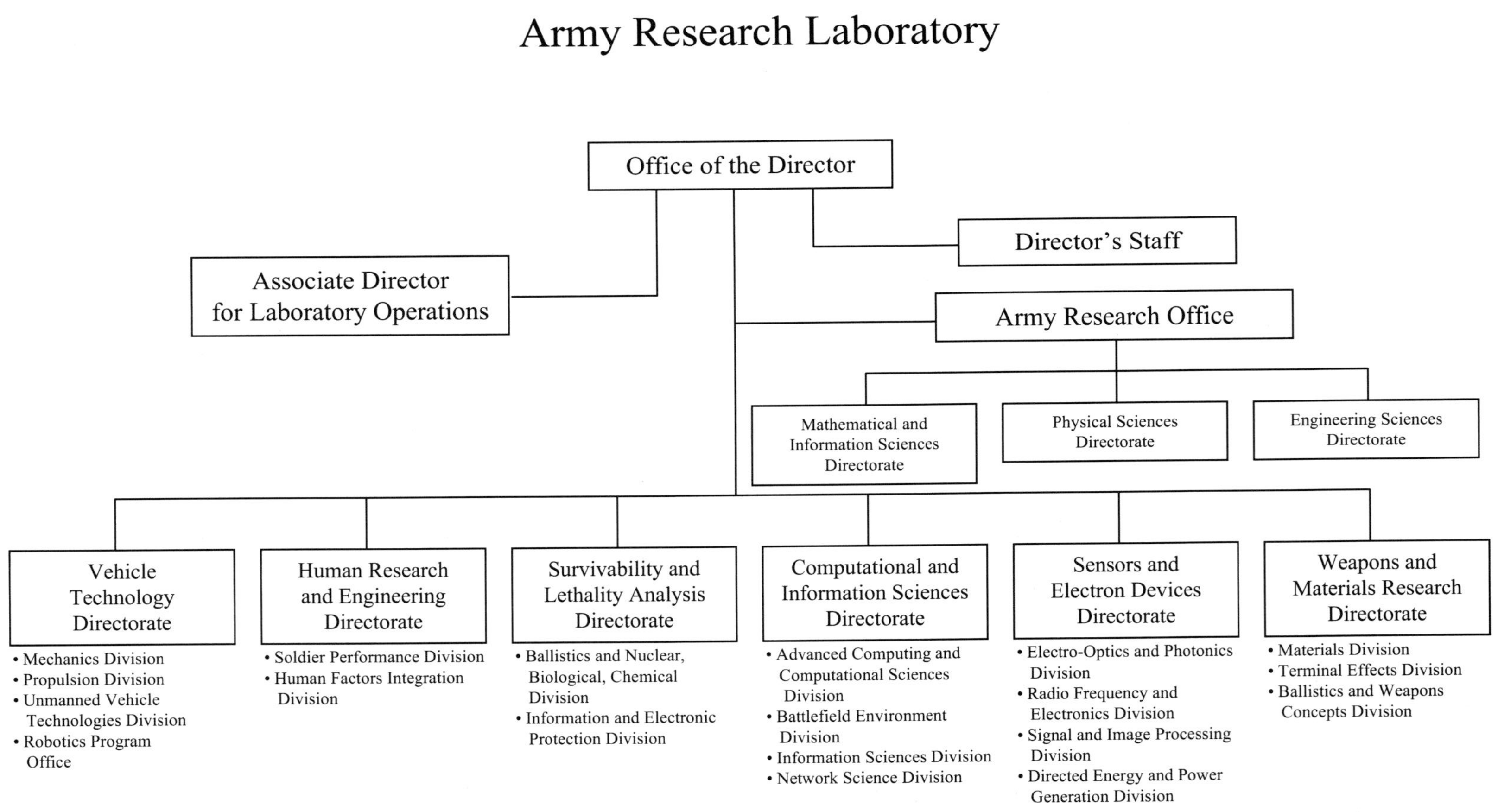

FIGURE A.1 Army Research Laboratory organization chart.

TABLE A.1 Resources: Army Research Laboratory Funding, by Directorate, for Fiscal Year 2003 Through Fiscal Year 2008 (millions of dollars)

		Technical Unit					
Type of Funding	Fiscal Year (FY)	CISD	HRED	SEDD	SLAD	VTD	WMRD
6.1	FY03	15.3	2.6	9.7	0.0	4.0	22.3
	FY04	13.2	2.6	20.5	0.0	3.5	20.0
	FY05	10.3	2.8	8.5	0.0	3.7	22.2
	FY06	9.6	2.7	8.4	0.0	3.7	23.3
	FY07	11.7	2.5	10.4	0.0	3.6	16.5
	FY08	13.0	2.9	12.6	0.1	4.0	15.9
6.1[a]	FY03	0.0	0.0	0.0	0.0	0.0	0.0
	FY04	0.0	0.0	0.0	0.0	0.0	0.0
	FY05	7.1	0.0	0.0	0.0	0.0	0.0
	FY06	9.8	0.0	0.0	0.0	0.0	0.0
	FY07	6.7	1.8	20.7	0.0	1.0	5.3
	FY08	7.8	0.9	12.6	0.0	0.0	9.8
6.1[b]	FY03	7.5	5.7	11.4	0.0	0.0	0.0
	FY04	7.9	6.0	11.9	0.0	0.0	2.4
	FY05	7.7	5.8	12.5	0.0	0.0	2.4
	FY06	11.6	5.4	9.8	0.0	0.0	2.4
	FY07	13.2	5.8	5.8	0.0	0.0	2.5
	FY08	13.7	5.4	5.7	0.0	3.0	0.0
6.1[c]	FY03	0.0	0.0	0.0	0.0	0.0	0.0
	FY04	0.0	0.0	0.0	0.0	0.0	0.0
	FY05	0.0	0.0	19.4	0.0	0.0	0.0
	FY06	0.0	0.0	16.7	0.0	0.0	0.0
	FY07	0.0	0.0	0.0	0.0	0.0	0.0
	FY08	0.0	0.0	0.0	0.0	0.0	0.0
6.2	FY03	17.9	19.6	51.4	6.8	4.5	83.1
	FY04	16.8	23.7	62.3	6.5	4.5	82.7
	FY05	13.5	16.7	56.7	7.0	4.7	88.2
	FY06	13.0	16.6	57.3	6.7	4.8	71.0
	FY07	13.3	18.9	51.9	6.7	5.4	65.4
	FY08	15.7	16.8	55.1	7.6	13.2	67.0
6.2[d]	FY03	0.0	0.0	0.0	0.0	0.0	0.0
	FY04	0.0	0.0	0.0	0.0	0.0	0.0
	FY05	4.2	0.0	0.0	0.0	0.0	0.0
	FY06	3.7	0.0	0.0	0.0	0.0	0.0
	FY07	5.2	21.8	33.1	0.0	1.2	47.4
	FY08	5.4	21.8	39.7	0.0	6.9	51.3
6.2[e]	FY03	0.0	0.0	0.0	0.0	0.0	5.4
	FY04	0.0	0.0	0.0	0.0	0.0	7.5
	FY05	0.0	0.0	0.0	0.0	0.0	8.0
	FY06	0.0	0.0	0.0	0.0	0.0	8.0
	FY07	0.0	0.0	3.9	0.0	0.0	7.1
	FY08	0.0	0.0	4.4	0.0	6.9	0.0

continued

TABLE A.1 Continued

Type of Funding	Fiscal Year (FY)	Technical Unit					
		CISD	HRED	SEDD	SLAD	VTD	WMRD
6.2[f]	FY03	0.0	0.0	0.0	0.0	0.0	0.0
	FY04	0.0	0.0	0.0	0.0	0.0	0.0
	FY05	0.0	0.0	41.1	0.0	0.0	0.0
	FY06	0.0	0.0	31.9	0.0	0.0	0.0
	FY07	0.0	0.0	0.0	0.0	0.0	0.0
	FY08	0.0	0.0	0.0	0.0	0.0	0.0
6.3/6.4/6.7	FY03	6.9	0.2	0.0	0.1	0.0	12.8
	FY04	0.0	0.0	7.2	1.0	0.0	14.5
	FY05	3.0	0.2	14.3	0.0	0.0	12.2
	FY06	0.0	0.7	17.8	0.0	0.0	20.8
	FY07	0.0	0.6	15.1	0.0	0.0	15.8
	FY08	0.0	0.6	10.4	0.0	0.0	32.5
6.3/6.4/6.7[g]	FY03	0.0	0.0	0.0	0.0	0.0	0.0
	FY04	0.0	0.0	0.0	0.0	0.0	0.0
	FY05	2.9	0.0	0.0	0.0	0.0	0.0
	FY06	0.0	0.0	0.0	0.0	0.0	0.0
	FY07	0.0	0.0	0.0	0.0	0.0	16.0
	FY08	0.0	0.0	0.0	0.0	0.0	9.2
6.6[h]	FY03	0.0	0.0	0.0	34.3	0.0	0.0
	FY04	0.0	0.0	0.0	40.2	0.0	0.0
	FY05	0.0	0.0	0.0	44.1	0.0	0.0
	FY06	0.0	0.0	0.0	40.0	0.0	0.0
	FY07	0.0	0.0	0.0	42.8	0.0	0.0
	FY08	0.0	0.0	0.0	39.1	0.0	0.0
6.6[i]	FY03	2.7	3.1	0.0	0.0	0.0	0.2
	FY04	0.1	3.0	0.0	0.0	0.0	0.0
	FY05	0.0	2.7	0.0	0.0	0.0	0.0
	FY06	0.0	2.5	0.0	0.0	0.0	0.0
	FY07	3.2	2.8	0.0	0.0	0.0	0.0
	FY08	5.2	1.8	0.0	1.6	0.0	0.0
6.6[j]	FY03	0.0	0.0	0.0	0.0	0.0	0.0
	FY04	0.0	0.0	0.0	0.0	0.0	0.0
	FY05	0.0	0.0	0.0	0.0	0.0	0.0
	FY06	0.0	0.0	0.0	0.0	0.0	0.0
	FY07	0.0	0.0	0.0	0.0	0.0	0.0
	FY08	0.0	0.0	0.0	0.0	0.0	0.0
Customer reimbursement[k]	FY03	28.6	7.9	22.7	15.4	1.0	45.5
	FY04	12.9	7.7	36.4	17.7	1.4	44.5
	FY05	17.2	8.9	78.2	71.1	1.2	63.3
	FY06	22.1	9.2	70.9	53.7	2.2	53.6
	FY07	39.5	11.5	55.2	25.3	3.1	82.9
	FY08	57.7	13.7	80.8	27.8	5.0	92.9

TABLE A.1 Continued

Type of Funding	Fiscal Year (FY)	Technical Unit					
		CISD	HRED	SEDD	SLAD	VTD	WMRD
Customer direct citation[l]	FY03	6.8	16.7	15.9	3.5	0.0	4.8
	FY04	5.3	0.3	17.8	10.3	0.0	5.7
	FY05	7.8	1.0	27.1	74.0	0.0	22.1
	FY06	10.3	1.0	81.7	41.5	0.0	20.1
	FY07	9.5	1.5	169.2	3.1	0.0	31.5
	FY08	11.9	2.0	375.7	4.6	0.0	7.7
OMA[m]	FY03	7.4	0.0	0.0	0.0	0.0	0.5
	FY04	0.6	0.0	0.0	0.0	0.0	1.9
	FY05	0.5	0.0	0.0	0.0	0.0	0.5
	FY06	0.5	0.0	24.6	0.0	0.0	0.6
	FY07	2.4	0.0	20.2	0.0	0.0	0.1
	FY08	0.5	0.7	0.0	0.0	0.0	0.6
OSD[n]	FY03	1.8	0.0	2.4	0.0	0.0	0.4
	FY04	18.8	0.0	0.0	0.0	0.0	0.0
	FY05	0.0	0.0	0.0	0.0	0.0	0.0
	FY06	0.4	0.0	23.6	0.0	0.0	0.0
	FY07	0.0	0.0	32.5	0.0	0.0	0.0
	FY08	0.0	0.0	6.6	0.0	0.0	2.1
DARPA[o]	FY03	1.5	0.1	44.9	0.2	0.0	2.9
	FY04	1.1	0.0	44.4	0.1	0.0	2.8
	FY05	2.4	0.8	75.7	0.0	0.0	2.0
	FY06	2.7	1.1	60.0	0.0	0.0	1.3
	FY07	3.5	0.2	37.7	0.0	0.0	1.7
	FY08	4.1	0.2	19.3	0.0	0.0	0.4
MSRC/HPC[p]	FY03	64.3	0.0	0.0	0.0	0.0	0.0
	FY04	57.7	0.0	0.0	0.0	0.0	0.0
	FY05	55.4	0.0	0.0	0.0	0.0	0.0
	FY06	53.5	0.0	0.0	0.0	0.0	0.0
	FY07	44.9	0.0	0.0	0.0	0.0	0.0
	FY08	60.0	0.0	0.0	0.0	0.0	0.0
MSRC/HPC[q]	FY03	0.0	0.0	0.0	0.0	0.0	0.0
	FY04	0.0	0.0	0.0	0.0	0.0	0.0
	FY05	15.0	0.0	0.0	0.0	0.0	0.0
	FY06	12.2	0.0	0.0	0.0	0.0	0.0
	FY07	9.4	0.0	0.0	0.0	0.0	0.0
	FY08	0.0	0.0	0.0	0.0	0.0	0.0
Total	FY03	160.7	55.9	158.4	60.3	9.5	177.9
	FY04	134.4	43.3	200.5	75.8	9.4	182.0
	FY05	147.0	38.9	333.5	196.2	9.6	220.9
	FY06	149.4	39.2	402.7	141.9	10.7	201.1
	FY07	162.5	67.4	455.7	77.9	14.3	292.2
	FY08	195.0	66.8	622.9	80.8	39.0	289.4

continued

TABLE A.1 Continued

NOTE: CISD, Computational and Information Sciences Directorate; HRED, Human Research and Engineering Directorate; SEDD, Sensors and Electron Devices Directorate; SLAD, Survivability and Lethality Analysis Directorate; VTD, Vehicle Technology Directorate; WMRD, Weapons and Materials Research Directorate.

[a] 6.1 Congressional funding.
[b] 6.1 Collaborative Technology Alliances. CISD FY06 includes the Network Sciences International Technology Alliance.
[c] 6.1 Collaborative Technology Alliances congressional funding.
[d] 6.2 Congressional funding.
[e] 6.2 Collaborative Technology Alliances.
[f] 6.2 Collaborative Technology Alliances congressional funding.
[g] 6.3/6.4/6.7 Congressional funding.
[h] 6.6 Technology analysis (SLAD, Small Business Innovation Research/Small Business Technology Transfer, Field Assistance in Science and Technology, Board of Army Science and Technology, Soldier Centered Analysis, and PE 65803 [Technical Information Activities]).
[i] 6.6 Congressional funding.
[j] 6.6 Management support (base support).
[k] Reimbursement from customers.
[l] Direct citation of funds from customers.
[m] Operation and Maintenance, Army.
[n] Office of the Secretary of Defense.
[o] Defense Advanced Research Projects Agency.
[p] Major Shared Resource Center and High Performance Computing (includes mission, OSD, and customer reimbursable).
[q] Major Shared Resource Center and High Performance Computing congressional funding.

SOURCE: Army Research Laboratory.

TABLE A.2 Army Research Laboratory Staffing Profiles, by Directorate, for the Years 2004, 2006, and 2008

Staffing Information		Number [%]					
		CISD	HRED	SEDD	SLAD	VTD	WMRD
Total civilian staff	Dec-04	313	217	376	294	84	403
	Jul-06	302	208	330	305	73	419
	Jul-08	294	186	391	305	73	402
Scientists and engineers	Dec-04	200 [64%]	162 [75%]	297 [79%]	236 [80%]	61 [73%]	282 [70%]
	Jul-06	88 [62%]	149 [72%]	272 [82%]	243 [80%]	50 [68%]	294 [70%]
	Jul-08	188 [64%]	137 [74%]	292 [75%]	245 [80%]	51 [70%]	288 [72%]
Technicians	Dec-04	16 [5%]	11 [5%]	46 [12%]	32 [11%]	11 [13%]	89 [22%]
	Jul-06	14 [5%]	16 [8%]	28 [9%]	32 [10%]	12 [17%]	89 [21%]
	Jul-08	21 [7%]	14 [8%]	65 [17%]	38 [12%]	14 [19%]	85 [21%]
Administrative personnel	Dec-04	97 [31%]	44 [20%]	33 [9%]	26 [9%]	12 [14%]	32 [8%]
	Jul-06	100 [33%]	43 [20%]	30 [9%]	30 [10%]	11 [15%]	36 [9%]
	Jul-08	85 [29%]	35 [19%]	34 [9%]	22 [7%]	8 [11%]	29 [7%]
Military personnel	Dec-04	6	3	4	15	5	5
	Jul-06	4	5	4	11	2	6
	Jul-08	5	6	3	9	3	2
Postdoctoral researchers	Dec-04	1	1	4	0	1	0
	Jul-06	0	1	17	0	0	8
	Jul-08	2	2	12	0	0	9
Guest researchers	Dec-04	2	3	11	0	0	0
	Jul-06	36	2	24	0	0	17
	Jul-08	12	5	10	0	3	10
On-site contractors	Dec-04	286	1	61	80	0	133
	Jul-06	260	3	65	123	0	214
	Jul-08	242	1	72	77	0	135
B.S. or B.A.	Dec-04	80 [40%]	64 [40%]	85 [29%]	136 [58%]	23 [38%]	101 [36%]
	Jul-06	58 [31%]	35 [23%]	52 [19%]	121 [50%]	9 [18%]	87 [30%]
	Jul-08	65 [35%]	40 [29%]	87 [30%]	134 [55%]	23 [45%]	89 [31%]
M.S. or M.A.	Dec-04	69 [35%]	51 [31%]	97 [32%]	78 [33%]	16 [26%]	62 [22%]
	Jul-06	81 [43%]	68 [46%]	99 [37%]	100 [41%]	21 [42%]	63 [21%][a]
	Jul-08	69 [37%]	41 [30%]	90 [31%]	85 [35%]	11 [22%]	66 [23%]
Ph.D.	Dec-04	51 [25%]	47 [29%]	115 [39%]	22 [9%]	22 [36%]	119 [42%]
	Jul-06	49 [26%]	46 [31%]	120 [44%]	22 [9%]	20 [40%]	144 [49%]
	Jul-08	49 [26%]	50 [36%]	111 [38%]	18 [7%]	16 [31%]	123 [43%]
Under 25 years of age	Dec-04	11 [6%]	14 [9%]	8 [3%]	19 [8%]	0	8 [3%]
	Jul-06	5 [3%]	7 [5%]	9 [3%]	12 [5%]	0	7 [2%]
	Jul-08	6 [3%]	1 [1%]	9 [3%]	11 [4%]	0 [0%]	0 [0%]

continued

TABLE A.2 Continued

Staffing Information		Number [%]					
		CISD	HRED	SEDD	SLAD	VTD	WMRD
25-35 years of age	Dec-04	19 [10%]	25 [15%]	33 [11%]	18 [8%]	7 [12%]	48 [17%]
	Jul-06	24 [13%]	25 [17%]	34 [13%]	39 [16%]	6 [8%]	39 [13%]
	Jul-08	27 [14%]	24 [18%]	48 [16%]	48 [20%]	5 [10%]	52 [18%]
35-45 years of age	Dec-04	63 [31%]	36 [22%]	112 [38%]	82 [35%]	28 [46%]	97 [34%]
	Jul-06	48 [26%]	35 [23%]	87 [32%]	72 [30%]	22 [30%]	98 [33%]
	Jul-08	35 [19%]	30 [22%]	60 [21%]	49 [20%]	11 [22%]	75 [26%]
45-55 years of age	Dec-04	65 [32%]	53 [33%]	81 [27%]	65 [27%]	14 [23%]	74 [26%]
	Jul-06	65 [35%]	46 [31%]	86 [32%]	76 [31%]	31 [43%]	78 [27%]
	Jul-08	73 [39%]	42 [31%]	113 [39%]	100 [41%]	23 [45%]	94 [33%]
55-65 years of age	Dec-04	36 [18%]	28 [17%]	52 [17%]	42 [18%]	10 [16%]	50 [18%]
	Jul-06	39 [21%]	30 [20%]	44 [16%]	36 [15%]	13 [18%]	66 [22%]
	Jul-08	37 [20%]	31 [23%]	48 [16%]	31 [13%]	11 [22%]	55 [19%]
Over 65 years of age	Dec-04	6 [3%]	6 [4%]	11 [4%]	10 [4%]	2 [3%]	5 [2%]
	Jul-06	7 [4%]	6 [4%]	12 [4%]	8 [3%]	1 [1%]	6 [2%]
	Jul-08	10 [5%]	8 [6%]	14 [5%]	6 [2%]	1 [2%]	12 [4%]

NOTE: CISD, Computational and Information Sciences Directorate; HRED, Human Research and Engineering Directorate; SEDD, Sensors and Electron Devices Directorate; SLAD, Survivability and Lethality Analysis Directorate; VTD, Vehicle Technology Directorate; WMRD, Weapons and Materials Research Directorate.

[a] Estimate.

SOURCE: Army Research Laboratory.

Appendix B

Membership of the Army Research Laboratory Technical Assessment Board and Its Panels

BIOGRAPHICAL SKETCHES OF MEMBERS: ARMY RESEARCH LABORATORY TECHNICAL ASSESSMENT BOARD

ROBERT W. BRODERSEN, *Chair,* is a member of the National Academy of Engineering, the John Whinnery Chair Professor in the Electrical Engineering and Computer Science Department at the University of California, Berkeley, and co-scientific director of the Berkeley Wireless Research Center. His expertise is in solid-state circuitry and microelectronics, and his current research is in new applications of integrated circuits focused on the areas of low-power design and wireless communications and the computer-aided design (CAD) tools necessary to support these activities. He is a fellow of the Institute of Electrical and Electronics Engineers (IEEE), and has received numerous prestigious awards throughout his career. Professor Brodersen received his Ph.D. in electrical engineering from the Massachusetts Institute of Technology.

DONALD B. CHAFFIN is a member of the National Academy of Engineering (NAE), the RG Snyder Distinguished University Professor, and the G. Lawton and Louise G. Johnson Professor of Industrial and Operations Engineering, Biomedical Engineering, and Occupational Health at the University of Michigan. He was elected into the NAE for fundamental engineering contributions to and leadership in occupational biomechanics and industrial ergonomics. Software resulting from his work is used in companies and universities throughout the world to evaluate people's risk of overexertion injuries when performing a variety of common manual tasks and to assist in designing workplaces and vehicles to better accommodate a diverse population. He is the founder and director of the Human Motion Simulation Laboratory at the University of Michigan. This laboratory is currently supported by GM, Ford Motor Company, Daimler Chrysler, International Truck and Engine Corporation, Lockheed Martin, the U.S. Postal Service, and the U.S. Army Tank-Automotive Command to develop and implement software modules to predict human motions and biomechanical limitations in CAD simulations that would affect the design of future vehicle and workplace systems. Dr. Chaffin has received numerous prestigious awards. He has published 105 peer-reviewed journal articles and 23 book chapters and co-authored 5 books, the latest entitled *Digital Human Modeling for Workplace and Vehicle Design.* He received his Ph.D. in industrial engineering from the University of Michigan.

PETER M. KOGGE is associate dean of engineering for research and also holds the McCourtney Chair in Computer Science and Engineering (CSE) at the University of Notre Dame. Prior to his joining Notre Dame in 1994, he was with IBM Federal Systems Division, and he was appointed an IEEE fellow in 1990 and an IBM fellow in 1993. In 1977, Dr. Kogge was a visiting professor in the Electrical and Computer Engineering Department at the University of Massachusetts, Amherst. From 1977 through 1994, he was also an adjunct professor in the Computer Science Department of the State University of New York at Binghamton. Since the summer of 1997, he has been a distinguished visiting scientist at the Center for Integrated Space Microsystems at the Jet Propulsion Laboratory. He is also the Research Thrust Leader for Architecture in Notre Dame's Center for Nano Science and Technology. For the 2000-2001 academic year, Dr. Kogge was the interim Schubmehl-Prein Chairman of the CSE Department at Notre Dame. Since the fall of 2003, he has also been a concurrent professor of electrical engineering. His research interests are in advanced computer architectures using unconventional technologies, such as processing-in-memory, and nanotechnologies, such as quantum-dot cellular automata.

KENNETH REIFSNIDER, a member of the National Academy of Engineering, is director of the Solid Oxide Fuel Program and professor of mechanical engineering at the University of South Carolina. Previously, he was Pratt and Whitney Chair Professor in Design and Reliability in the Department of

Mechanical Engineering at the University of Connecticut and director, Connecticut Global Fuel Cell Center. His research areas include applied mechanics, prediction of the lifetime of materials and structures, advanced materials, and fuel cells. Dr. Reifsnider joined the Mechanical Engineering Department at the University of Connecticut in 2002 from the Virginia Polytechnic Institute and State University, where he was the Alexander Giaco Chair Professor of Engineering Science and Mechanics and where he began the Virginia Tech Center for Composite Materials and Structures and served as director of the Virginia Institute for Material Systems. He also served as deputy director of the National Science Foundation Center for High Performance Polymeric Adhesives and Composites. In addition, he served as chair of the Materials Engineering Science Ph.D. Program and as associate provost for interdisciplinary programs at Virginia Tech. Dr. Reifsnider has received many prestigious awards throughout his career, serves on the editorial boards of five journals, is editor-in-chief of the *International Journal of Fatigue,* and is co-founding editor of the *Journal of Composites Technology and Research.* He also recently completed his signature text entitled *Damage Tolerance and Durability of Material Systems*. Dr. Reifsnider earned his Ph.D. in metallurgy and solid mechanics from the Johns Hopkins University. He is a fellow of the American Society of Mechanical Engineers.

JOHN C. SOMMERER is director of science and technology and chief technology officer of the Johns Hopkins University Applied Physics Laboratory (JHU/APL). He manages APL's overall research and development program and oversees APL's technology transfer program; he also oversees the participation of APL in JHU educational programs and serves as primary technical liaison with the academic divisions of the university. In addition, he is an adjunct faculty member in applied physics, applied mathematics, and technical management. Dr. Sommerer has made internationally recognized theoretical and experimental contributions to the fields of nonlinear dynamics and complex systems. He has served on several technical advisory bodies for the U.S. government, including a recent assignment as vice chair of the Naval Research Advisory Council, the senior technical advisory body to the Secretary of the Navy, the Chief of Naval Operations, and the Commandant of the Marine Corps. He holds a Ph.D. in physics from the University of Maryland.

DWIGHT C. STREIT is a member of the National Academy of Engineering and vice president for foundation technologies at Northrop Grumman Space Technology. He has overall responsibility for the development of the basic engineering, science, and technology required for space and communications systems. He has extensive experience in semiconductor devices and Monolithic Microwave Integrated Circuits for applications up to 220 gigahertz, as well as in infrared and radiometer sensors. He has led development efforts for 10 to 40 gigabit per second optical communication systems, and has experience in the development and production of optoelectronic devices and circuits. He also has previous experience in frequency-modulated continuous wave and phased-array product development for X-band to W-band radar applications. He received his Ph.D. in electrical engineering from the University of California, Los Angeles, in 1986.

Staff

JAMES P. McGEE is the director of the Laboratory Assessments Board, the Army Research Laboratory Technical Assessment Board (ARLTAB), and the Committee on National Institute of Standards and Technology Technical Programs, within the Division on Engineering and Physical Sciences at the National Research Council (NRC). Since 1994, he has been a senior staff officer at the NRC, directing projects in the areas of systems engineering and applied psychology, including activities of ARLTAB and projects of the Committee on National Statistics' (CNS's) Panel on Operational Testing and Evaluation of the Stryker Vehicle and CNS's Committee on Assessing the National Science Foundation's Scientists and Engineers Statistical Data System; the Committee on the Health and Safety Needs of Older Workers; and the Steering Committee on Differential Susceptibility of Older Persons to Environmental Hazards. He has also served as staff officer for NRC projects on Air Traffic Control Automation, Musculoskeletal Disorders and the Workplace, and the Changing Nature of Work. Prior to joining the NRC, Dr. McGee held technical and management positions in systems engineering and applied psychology at IBM, General Electric, RCA, General Dynamics, and United Technologies corporations. He received his B.A. from Princeton University and his Ph.D. from Fordham University, both in psychology, and for several years instructed postsecondary courses in applied psychology and in organizational management.

ARUL MOZHI is senior program officer at the Laboratory Assessments Board within the Division on Engineering and Physical Sciences at the National Research Council (NRC). Since 1999, he has been a senior program officer at the NRC, directing projects in the areas of defense science and technology, including those carried out by numerous study committees of the Laboratory Assessments Board, the Army Research Laboratory Technical Assessment Board, the Naval Studies Board, the National Materials Advisory Board, and the Board on Manufacturing and Engineering Design. Prior to joining the NRC, Dr. Mozhi held technical and management positions in systems engineering and applied materials research and development at UTRON, Inc.; Roy F. Weston, Inc.; and Marko Materials, Inc. He received his M.S. and Ph.D. degrees (the latter in 1986) in materials engineering from the Ohio State University and then served as a postdoctoral research associate there. He received his B.S. in metallurgical engineering from the Indian Institute of Technology in 1982.

LIZA HAMILTON is the administrative coordinator for the Laboratory Assessments Board within the Division on Engineering and Physical Sciences at the National Research Council (NRC). Since 2002, she has been responsible for managing the administrative aspects of panel formation, panel meetings, report publication and dissemination, and program development. In addition, she has designed newsletters and has rendered cover designs and figures for numerous reports prepared by the NRC's Division on Life Sciences and Division on Engineering and Physical Sciences. Ms. Hamilton earned a B.F.A. in film studies from the University of Utah and a design certification from Maryland Institute College of Art. She is currently completing an M.L.A. from the Johns Hopkins University.

PANEL ROSTERS

Panel on Air and Ground Vehicle Technology

Kenneth Reifsnider, University of South Carolina, *Chair*
Ralph Aldredge, University of California, Davis
Meyer Benzakein, Ohio State University
James Bettner, Propulsion Consultant, Pittsboro, Indiana
Paul Bevilaqua, Lockheed Martin Aeronautics Company
Julie Chen, University of Massachusetts, Lowell
David Crow, Pratt and Whitney (retired)
Earl Dowell, Duke University
S. Michael Hudson, Rolls-Royce North American Technologies, Inc. (retired)
William McCroskey, NASA Ames Research Center
Thomas Mueller, University of Notre Dame
Lynne Parker, University of Tennessee
Neil Paton, Liquidmetal Technologies
Martin Peryea, Bell Helicopter Textron, Inc.
William Sirignano, University of California, Irvine
Christine Sloane, General Motors Corporation
Michael Torok, Sikorsky Aircraft Corporation
James Williams, Ohio State University
Ronald York, Rolls-Royce North American Technologies, Inc.

Panel on Armor and Armaments

Kim Baldridge, University of Zurich
Thomas Brill, University of Delaware
Thomas Eagar, Massachusetts Institute of Technology
Mark Eberhart, Colorado School of Mines
Richard Farris, University of Massachusetts, Amherst
Katharine Frase, IBM Corporation
George (Rusty) Gray III, Los Alamos National Laboratory
Rigoberto Hernandez, Georgia Institute of Technology
K. Sharvan Kumar, Brown University
R. Bowen Loftin, Texas A&M University
Gregory Miller, University of California, Davis
Anita Renlund, Sandia National Laboratories
Christopher Schuh, Massachusetts Institute of Technology
Leonard Uitenham, North Carolina Agricultural and Technical State University

Panel on Digitization and Communications Science

Peter Kogge, University of Notre Dame, *Chair*
Mikhail Atallah, Purdue University
Steven Bellovin, Columbia University
Willard Bolton, Sandia National Laboratories

Robert Brodersen, University of California, Berkeley
L. Reginald Brothers, Jr., BAE Systems
Gary Brown, Virginia Polytechnic Institute and State University
Lori Freitag Diachin, Lawrence Livermore National Laboratory
Joel Engel, JSE Consulting, Armonk, New York
William Gropp, University of Illinois at Urbana-Champagne
Robert Lucas, University of Southern California
Jimmy Omura, Gordon and Betty Moore Foundation
Tamar Peli, The Charles Stark Draper Laboratory, Inc.
Mikel Petty, University of Alabama, Huntsville
John Snow, University of Oklahoma
David Waltz, Columbia University

Panel on Sensors and Electron Devices

Dwight Streit, Northrop Grumman Space Technology, *Chair*
Ilesanmi Adesida, University of Illinois at Urbana-Champaign
Donald Chiarulli, University of Pittsburgh
J. Patrick Fitch, National Biodefense Analysis and Countermeasures Center
Daniel Fuhrmann, Washington University, St. Louis
Thomas Fuller, Georgia Institute of Technology
Herbert Hess, University of Idaho
Paul Hoff, Independent Consultant, Bedford, New Hampshire
Leslie Kolodziejski, Massachusetts Institute of Technology
Douglas Mook, The Aptec Group
Albert Pisano, University of California, Berkeley
Zoya Popovic, University of Colorado, Boulder
P. Paul Ruden, University of Minnesota
James Sabatier, University of Mississippi
Edmund Schweitzer III, Schweitzer Engineering Laboratories, Inc.
Subhash Singhal, Pacific Northwest National Laboratory
Levi Thompson, University of Michigan
Steven Visco, Lawrence Berkeley National Laboratory

Panel on Survivability and Lethality Analysis

John Sommerer, Johns Hopkins University Applied Physics Laboratory, *Chair*
David Aucsmith, Microsoft Corporation
David Barton, Independent Consultant, Hanover, New Hampshire
Thomas Burris, Lockheed Martin Aeronautics Company
MarjorieAnn EricksonKirk, Phoenix Engineering Associates, Inc.
Alan Jones, The Boeing Company
Hilarie Orman, Purple Streak, Inc.
Tibor Schonfeld, Johns Hopkins University Applied Physics Laboratory
Donald Wunsch, Printron, Inc.

Soldier Systems Panel

Donald Chaffin, University of Michigan, *Chair*
Julie Adams, Vanderbilt University
Theodore Berger, University of Southern California
Tora Bikson, The RAND Corporation
Michael Byrne, Rice University
Steven Hyman, Harvard University
Daniel Ilgen, Michigan State University
Gerald Krueger, Krueger Ergonomics Consultants, Vienna, Virginia
Michael Merzenich, University of California, San Francisco
Virginia Richards, University of Pennsylvania
Emilie Roth, Roth Cognitive Engineering
Gavriel Salvendy, Purdue University
Thomas Sanquist, Pacific Northwest National Laboratory
Deborah Thompson, BAE Systems
Richard Thompson, University of Southern California
Leslie Ungerleider, National Institutes of Health
Joel Warm, University of Cincinnati
Jeremy Wolfe, Harvard University
Michael Zyda, GamePipe Laboratory

Appendix C

Assessment Criteria

The Army Research Laboratory Technical Assessment Board's assessment considered the following general questions posed by the ARL Director:

1. Is the scientific quality of the research of comparable technical quality to that executed in leading federal, university, and/or industrial laboratories both nationally and internationally?
2. Does the research program reflect a broad understanding of the underlying science and research conducted elsewhere?
3. Does the research employ the appropriate laboratory equipment and/or numerical models?
4. Are the qualifications of the research team compatible with the research challenge?
5. Are the facilities and laboratory equipment state of the art?
6. Does the research reflect an understanding of the Army's requirement for the research or the analysis?
7. Are programs crafted to employ the appropriate mix of theory, computation, and experimentation?
8. Is the work sufficiently unique and appropriate to the ARL niche?
9. Are there especially promising projects that, with application of adequate resources, could produce outstanding results that could be transitioned ultimately to the field?

The Board applied the following metrics or criteria to the assessment of the scientific and technical work reviewed at the Army Research Laboratory (ARL):

1. Effectiveness of Interaction with the Scientific and Technical Community
   a. Papers in quality refereed journals and conference proceedings (and their citation index)
   b. Presentations and colloquia

   c. Participation in professional activities (society officers, conference committees, journal editors)
   d. Educational outreach (serving on graduate committees, teaching or lecturing, invited talks, mentoring students)
   e. Fellowships and awards (external and internal)
   f. Review panel participation (Army Research Office, National Science Foundation, Multidisciplinary University Research Imitative)
   g. Recruiting new talent into the ARL
   h. Patents and intellectual property (IP) (and examples of how the patent or IP is used)
   i. Involvement in building an ARL-wide cross-directorate community
   j. Public recognition (e.g., in the press and elsewhere) for ARL research

2. Impact on Customers
   a. Documented transfer or transition of technology, concepts, or program assistance from ARL to Research, Development, and Engineering Centers (RDECs) or RDEC contractors for both the long term and short term
   b. Direct funding from customers to support ARL activities
   c. Documented demand for ARL support or services (is there competition for ARL's support?)
   d. Customer involvement in directorate planning
   e. Participation in multidisciplinary, cross-directorate projects
   f. Surveys of customer base (direct information from customers on value of ARL research)

3. Formulation of Projects' Goals and Plans
   a. Is there a clear tie to ARL Strategic Focus Areas, Strategic Plan, or other ARL need?
   b. Are tasks well defined to achieve objectives?
   c. Does the project plan clearly identify dependencies (i.e., successes depend on success of other activities within the project or outside developments)?
   d. If the project is part of a wider activity, is role of the investigators clear, and are the project tasks and objectives clearly linked to those of other related projects?
   e. Are milestones identified if they are appropriate? Do they appear feasible?
   f. Are obstacles and challenges defined (technical, resources)?
   g. Does the project represent an area where application of ARL strengths is appropriate?

4. Research and Development Methodology
   a. Are the hypotheses appropriately framed within the literature and theoretical context?
   b. Is there a clearly identified and appropriate process for performing required analyses, prototypes, models, simulations, tests, etc.?
   c. Are the methods (e.g., laboratory experiment, modeling or simulation, field test, analysis) appropriate to the problems? Do these methods integrate?
   d. Is the choice of equipment or apparatus appropriate?
   e. Is the data collection and analysis methodology appropriate?
   f. Are conclusions supported by the results?
   g. Are proposed ideas for further study reasonable?
   h. Do the trade-offs between risk and potential gain appear reasonable?
   i. If the project demands technological or technical innovation, is that occurring?
   j. What stopping rules, if any, are being or should be applied?

5. Capabilities and Resources
    a. Are the qualifications and number of the staff (scientific, technical, administrative) appropriate to achieve success of the project?
    b. Is funding adequate to achieve success of the project?
    c. Is the state of the equipment and facilities adequate?
    d. If staff, funding, or equipment is not adequate, how might the project be triaged (what thrust should be emphasized, what sacrificed?) to best move toward its stated objectives?
    e. Does the laboratory sustain the technical capability to respond quickly to critical issues as they arise?

6. Responsiveness to the Board's Recommendations
    a. Have the issues and recommendations presented in the previous report been addressed?

Appendix D

Acronyms

AC&CD	Advanced Computing and Computational Division
AC&CSD	Advanced Computing and Computational Sciences Division
ACAT	Air Coupled Thermography Inspection
ACSAC	Applied Computer Security Applications Conference
AFME	Armed Forces Medical Examiner
AFV	armored fighting vehicles
AHPCRC	Army High Performance Computing Research Center
ALEGRA	Arbitrary-Lagrangian-Eulerian General Research Applications
AMRDEC	Aviation and Missile Research, Development and Engineering Center
APD	avalanche photodiode
APG	Aberdeen Proving Ground
APS	active protection system
ARDEC	Armament Research, Development and Engineering Center
ARL	Army Research Laboratory
ARLTAB	Army Research Laboratory Technical Assessment Board
ARO	Army Research Office
ASA	Acoustical Society of America
ASCED	Active Stall Control Engine Demonstration
ASO	Army Science Objective
ASU	Arizona State University
ATO	Army Technology Objective
BAA	Broad Agency Announcement
BED	Battlefield Environment Division

BRAC	base realignment and closure
C4ISR	command, control, communications, computers, intelligence, surveillance, and reconnaissance
CAD	computer-aided design
CAN	computer network attacks
CAT-ATD	Crew-Integration and Automation Testbed Advanced Technology Demonstrator
CBM	condition-based maintenance
CECOM	Communications and Electronics Command Center
CERDEC	Communications-Electronics Research, Development, and Engineering Center
CFD	computational fluid dynamics
CIO/G-6	Office of the Chief Information Officer
CISD	Computational and Information Sciences Directorate
CMS	Computational Materials Science
CNE	computer network exploitation
CPU	central processing unit
C-QWIP	corrugated quantum-well infrared photodetector
CRADA	Cooperative Research and Development Agreement
CSEB	Computational Sciences and Engineering Branch
CSG	constructive solid geometry
CTA	Collaborative Technology Alliance
DARPA	Defense Advanced Research Projects Agency
DFT	density functional theory
DNA	deoxyribonucleic acid
DoD	Department of Defense
DOE	Department of Energy
DRC	Dynamics Research Corporation
DRM	digital rights management
DSRC	DoD Supercomputing Resource Center
DTRA	Defense Threat Reduction Agency
DU	depleted uranium
EAR	Environment for Auditory Research
ECAE	equal channel angle extrusion
EDM	Engineering Development Model
EFP	explosively formed penetrator
EM	electromagnetic
EOS	equation of state
EW	electronic warfare
FCS	Future Combat Systems
FDC	Flexible Display Center
FEM	finite element modeling
FET	field-effect transistor
FIB	focused ion beam

FM	frequency modulation, frequency modulated
FPGA	field-programmable gate array
FY	fiscal year
GPS	Global Positioning System
GTO	gate turn-off thyristor
HEMT	high electron mobility transistor
HF	human factors; hydrofluoric acid
HPC	high-performance computing
HRED	Human Research and Engineering Directorate
HSARPA	Homeland Security Advanced Research Projects Agency
HUMS	health and usage monitoring
I2WD	Intelligence Information Warfare Directorate
IA	information assurance
ICE	IED counter electronic (device)
IED	improvised explosive device
IM	insensitive munitions
IMPRINT	Improved Performance Research Integration Tool (software)
INL	Idaho National Laboratory
IO	information operations
IOC	initial operational capability
IOL	Intelligent Optics Laboratory
IP	Internet Protocol
IR	infrared
ISD	Information Sciences Division
ISR	intelligence, surveillance, and reconnaissance
ITA	International Technology Alliance
JTAPIC	Joint Trauma Analysis and Prevention of Injury in Combat
JTRS	Joint Tactical Radio System
KE	kinetic energy
LWIR	long-wavelength infrared
MANET	mobile ad hoc network
MANPRINT	Manpower and Personnel Integration
MASINT	Measurement and Signature Intelligence
MBE	molecular-beam epitaxy
MCoE	Materials Center of Excellence
MCS	mounted combat system
MDD	multidisciplinary design
MEMS	microelectromechanical systems
MMF	Mission and Means Framework (software)

MMW	millimeter-wave
MOCVD	metal-organic chemical vapor deposition
MOSFET	metal-oxide-semiconductor field-effect transistor
MRAP	mine-resistant ambush-protected
MRB	Management Review Board
MSS BAMS	Military Sensing Symposium on Battlefield Acoustic and Magnetic Sensing
MT	Machine Translation (Program)
MTI	moving target indicator
MURI	Multidisciplinary University Research Initiative
MUVES	Modular UNIX-based Vulnerability Estimation Suite (software)
MWIR	mid-wavelength infrared
NAMD	NAnoscale Molecular Dynamics
NASA	National Aeronautics and Space Administration
NBC	nuclear, biological, and chemical
NCI	nanoscale-compositional-inhomogenous
NDE	nondestructive evaluation
NERF	Novel Energetic Research Facility
NGIC	National Ground Intelligence Center
NMSU	New Mexico State University
NRC	National Research Council
NSD	Network Science Division
NURB	non-uniform rational B-spline surfaces
OFDM	orthogonal frequency division multiplexing
ORCA	Operational Requirements-based Casualty Assessment
PARC	Palo Alto Research Center
PDA	personal digital assistant
PI	principal investigator
PKI	public key infrastructure
PUNDIT	Parallel Unsteady Domain Information Transfer
PZT	lead zirconium titanate
QE	quantum efficiencies
QWIP	AlGaAs/GaAs quantum-well inter-subband photodetector
R&D	research and development
RA	reactive armor
RDEC	Research, Development, and Engineering Center
RDECOM	Research, Development, and Engineering Command
RDT&E	research, development, test and evaluation
RF	radio frequency
RM	reactive material
S&Es	scientists and engineers

S4	Systems of Systems Survivability Simulation (software)
SAR	synthetic aperture radar
SBIR	Small Business Innovation Research
SEDD	Sensors and Electron Devices Directorate
SEM	systems effective modeling
SINCGARS	single channel ground and airborne radio system
SLAD	Survivability and Lethality Analysis Directorate
SLV	survivability, lethality, and vulnerability
SoS	system of systems
SRW	soldier radio waveform
STAR	Scalable Technology for Adaptive Response
STI	Strategic Technology Initiative
SWFTICE	Sensor, Warhead and Fuze Technology Integrated for Combined Effects
TARDEC	Tank and Automotive Research, Development, and Engineering Center
TBI	traumatic brain injury
TFT	thin-film transistor
THINK	Tactical Human Integration with Networked Knowledge
TRADOC	Training and Doctrine Command
TWV	tactical wheeled vehicle
UAV	unmanned aerial vehicle
UGS	unattended ground sensor
UGV	unmanned ground vehicle
UHF	ultrahigh frequency
UHMWPE	ultra high molecular weight polyethylene
UML	unified modeling language
USC ISI	University of Southern California's Information Sciences Institute
UTAMS	unattended transient acoustic MASINT sensor
UV	ultraviolet
UWB	ultrawideband
V&V	validation and verification (efforts)
VLWIR	very long wavelength infrared
VTD	Vehicle Technology Directorate
WMRD	Weapons and Materials Research Directorate
WNW	wideband networking waveform
WSMR	White Sands Missile Range